BRITISH PUFFBALLS EARTHSTARS AND STINKHORNS

BRITISH PUFFBALLS, EARTHSTARS AND STINKHORNS

AN ACCOUNT OF THE BRITISH GASTEROID FUNGI

D.N. PEGLER

T. LÆSSØE

B.M. SPOONER

Royal Botanic Gardens, Kew

© Copyright The Board of Trustees of The Royal Botanic Gardens, Kew 1995

First published 1995

Typeset at the Royal Botanic Gardens, Kew

Cover Design by Media Resources, RBG, Kew

ISBN 0 947643 81 8

Printed and Bound in Great Britain by Whitstable Litho, Whitstable, Kent

Dedicated to Mr. J. Terence Palmer, who is recognized throughout the world as a leading expert on the gasteroid fungi. The following account owes much to his *Chronological Catalogue of the Literature to the British Gasteromycetes, 1968.*

CONTENTS

Introduction 1
Form and Structure 3
Distribution and Habitats 10
How to Collect and Examine Gasteroid Fungi 13
Folklore and Uses 14
Keys to Orders and Families 19

Earthballs, Dyeball and Barometer Earthstar (Sclerodermatales), key to families 20
  Sclerodermataceae, key to genera 21
    Earthballs (*Scleroderma*), key to species 22
      Potato Earthball (*Scleroderma bovista*) 24
      Common Earthball (*Scleroderma citrinum*) 26
      Many-rooted Earthball (*Scleroderma polyrhizum*) 28
      Onion Earthball (*Scleroderma cepa*) 30
      Scaly Earthball (*Scleroderma verrucosum*) 32
      Leopard-spotted Earthball (*Scleroderma areolatum*) 34
    Dyeball (*Pisolithus arhizus*) 36
  Astraeaceae 38
    Barometer Earthstar (*Astraeus hygrometricus*) 40

Stalk and Stilt Puffballs (Tulostomatales), key to families 42
  Tulostomataceae; key to genera 42
    Stalk Puffballs (*Tulostoma*), key to species 43
      White Stalk Puffball (*Tulostoma niveum*) 44
      Winter Stalk Puffball (*Tulostoma brumale*) 46
      Scaly Stalk Puffball (*Tulostoma melanocyclum*) 48
      Quélet's Puffball (*Queletia mirabilis*) 50
    Stilt Puffballs (Battarraeaceae) 52
      Sandy Stilt Puffball (*Battarraea phalloides*) 53

Bird's Nest Fungi and Cannon Fungus (Nidulariales), key to families 56
  Nidulariaceae, key to genera 56
    Bird's Nest Fungi (*Cyathus*), key to species 57
      Fluted Bird's Nest (*Cyathus striatus*) 58
      Dung Bird's Nest (*Cyathus stercoreus*) 60
      Field Bird's Nest (*Cyathus olla*) 62
      White-egg Bird's Nest (*Crucibulum, C. laeve*) 64
      Pea-shaped Bird's Nest (*Nidularia, N. deformis*) 66
    Mycocalia, key to species 68
    Bog Mycocalia (*Mycocalia sphagneti*) 69
    Tiny Mycocalia (*Mycocalia minutissima*) 70
    Common Mycocalia (*Mycocalia denudata*) 71
    Durieu's Mycocalia (*Mycocalia duriaeana*) 72
    Cannon Fungus (Sphaerobolaceae, *Sphaerobolus stellatus*) 74

Puffballs, Bovists and Earthstars (Lycoperdales), key to families **76**

Geastraceae, key to genera 76

Earthstars (*Geastrum*), key to species 77

Elegant Earthstar (*Geastrum elegans*) 80

Field Earthstar (*Geastrum campestre*) 82

Berkeley's Earthstar (*Geastrum berkeleyi*) 84

Dwarf Earthstar (*Geastrum schmidelii*) 86

Striated Earthstar (*Geastrum striatum*) 88

Beaked Earthstar (*Geastrum pectinatum*) 90

Daisy Earthstar (*Geastrum floriforme*) 92

Sessile Earthstar (*Geastrum fimbriatum*) 94

Arched Earthstar (*Geastrum fornicatum*) 96

Four-rayed Earthstar (*Geastrum quadrifidum*) 98

Tiny Earthstar (*Geastrum minimum*) 100

Rosy Earthstar (*Geastrum rufescens*) 102

Crowned Earthstar (*Geastrum coronatum*) 104

Weather Earthstar (*Geastrum corollinum*) 106

Collared Earthstar (*Geastrum triplex*) 108

Flask-shaped Earthstar (*Geastrum lageniforme*) 110

Pepper Pot (*Myriostoma, M. coliforme*) 112

Lycoperdaceae, key to genera 114

Giant Puffball (*Calvatia, C. gigantea*) 116

Meadow Puffball (*Vascellum, V. pratense*) 118

Rooting Bovist (*Bovistella, B. radicata*) 120

Handkea, key to species 122

Mosaic Puffball (*Handkea utriformis*) 124

Pestle-shaped Puffball (*Handkea excipuliformis*) 126

Bovista, key to species 128

Deceiving Bovist (*Bovista aestivalis*) 130

Dwarf Bovist (*Bovista dermoxantha*) 132

Fen Bovist (*Bovista paludosa*) 134

Least Bovist (*Bovista limosa*) 136

Brown Bovist (*Bovista nigresecens*) 138

Lead-grey Bovist (*Bovista plumbea*) 140

Lycoperdon, key to species 142

Stump Puffball (*Lycoperdon pyriforme*) 144

Flaky Puffball (*Lycoperdon mammiforme*) 146

Pedicelled Puffball (*Lycoperdon caudatum*) 148

Hedgehog Puffball (*Lycoperdon echinatum*) 150

Common Puffball (*Lycoperdon perlatum*) 152

Blackish Puffball (*Lycoperdon nigrescens*) 154

Grassland Puffball (*Lycoperdon lividum*) 156

Dark-spored Puffball (*Lycoperdon atropurpureum*) 158

Soft-spined Puffball (*Lycoperdon molle*) 160

Steppe Puffball (*Lycoperdon decipiens*) 162

Umber Puffball (*Lycoperdon umbrinum*) 164

Conifer Puffball (*Lycoperdon lambinonii*) 166

Heather Puffball (*Lycoperdon ericaeum*) 168

Stinkhorns & Cage Fungi (Phallales), key to families **170**

Phallaceae, key to genera 171

Stinkhorns (*Phallus*), key to species 171

Common Stinkhorn (*Phallus impudicus* and var. *togatus*) 172

Dune Stinkhorn (*Phallus hadriani*) 174

Mutinus, key to species 176

Dog Stinkhorn (*Mutinus caninus*) 177

Red Stinkhorn (*Mutinus ravenelii*) 178

Clathraceae, key to genera 180

Cage Fungi (*Clathrus*), key to species 181

Devil's Fingers (*Clathrus archeri*) 182

Red Cage Fungus (*Clathrus ruber*) 184

Lizard's Claw (*Lysurus cruciatus*) 186

Starfish Fungus (*Aseroe rubra*) 188

Basket Fungus (*Ileodictyon cibarium*) 190

Keys to Hypogeous Gasteroid Fungi (False Truffles) **193**

Glossary **199**

References **203**

BMS Distribution Maps **213**

Index **251**

INTRODUCTION

The fungi considered in this book were, until recently, classified together as a single group, *Gasteromycetes*, forming a class of *Basidiomycotina*. They have in common the fact that the spores are developed within closed fruitbodies, a so-called angiocarpic development, one result of which is that the spores are no longer forcibly discharged from the basidium. Instead, they are passively released, and dispersed with the help of external agents such as wind, raindrops or insects. It is now realized that these fungi represent an heterogeneous assemblage, a mixture of forms which are derived from various lineages and which cannot be regarded as forming a single taxonomic unit. These fungi can be collectively referred to as 'gasteroid fungi', all being similar in having a closed development, but they cannot be classified together as a single group. The class *Gasteromycetes* is clearly polyphyletic. It must, therefore, be abandoned and the fungi referred there reclassified, placed elsewhere in the system to reflect their true relationships with other groups. However, due to the long tradition which exists of treating these fungi as representatives of a single group, it is still convenient to consider them together in a single volume.

The gasteroid fungi exhibit a wide range of form and structure. Many such fungi have a hypogeous development, their fruitbodies developing and maturing below the soil surface or buried in leaf litter. These fungi are sometimes known as false truffles. They are not closely related to the gasteroid fungi with epigeous development and are not included in this book. A revision of the British hypogeous species was published by Pegler, Spooner & Young (1993).

Gasteroid fungi which are epigeous at maturity can mostly be classified into five orders. Three of these, *Sclerodermatales, Lycoperdales* and *Tulostomatales* are puffball-like, developing at maturity a powdery spore mass which is dispersed by wind or water. Another, *Nidulariales*, includes the bird's-nest fungi, in which spores are developed in 'eggs' or peridioles and do not form a powdery mass. The Phallales includes the stinkhorns, cage fungi and their allies, soft-fleshed fungi which develop from a gelatinous 'egg-stage', and produce spores in a slimy, unpleasant-smelling mass designed for dispersal by flies. In addition to these are several little-known gasteroid fungi which are of varied or uncertain affinity, and which are not included in the present work. British examples of these include the marine white rot fungus *Nia vibrissa* Moore & Meyers (*Melanogastraceae*) (see Leightley & Eaton 1979), and *Mycaureola dilseae* Maire & Chemin, which is parasitic on marine red algae and whose taxonomic position is as yet unknown (see Porter & Farnham 1986). A number of gasteroid 'heterobasidiomycetes' (fungi with septate basidia) are also known, some of which have been reported from Britain. These include *Heterogastridium* Oberw. & Bauer, recently described in the new monotypic order *Heterogastridiales* (Oberwinkler & Bauer 1990), and the auricularioid species *Atractiella solani* (Cohn & Schröt.) Oberw. & Bandoni and *Phloeogena faginea* (Fr.) Link (*Atractiellales*), *Stilbum vulgare* Tode (*Chionosphaeraceae*) and *Pachnocybe ferruginea* (*Pachnocybaceae*) (see Oberwinkler & Bauer 1989).

HISTORICAL OUTLINE

Although popular interest in gasteroid fungi in the British Isles has a long history, there was little scientific study of the British species until the latter part of the 18th century. The earliest accounts were mainly descriptive synopses of the known species, or floristic studies such as those of Hudson (1762), Curtis (1772 - 1798) and Withering

(1776). Bolton (1788) provided descriptions as well as good coloured illustrations of those known from the Halifax region. However, Bryant (1782) provided an historical account of two species of *Lycoperdon*, and the first critical account of British earthstars was published by Woodward (1794). The 19th century saw much more progress in the study of these fungi, particularly from workers such as Berkeley, Broome, Cooke, who published a complete list of British fungi in 1863 (Cooke 1863), Currey, Higgins, Massee, Plowright, Phillips, and W.G.Smith. In 1889, Massee published 'A Monograph of the British Gastromycetes'. This was an illustrated account covering all that was known of the British species, including the subterranean forms (false truffles). It was a major contribution to the study of the British gasteroid fungi, greatly expanding an earlier account by Higgins (1859). Since then, the only other descriptive account of this group in its entirety is to be found in Carlton Rea's 'British Basidiomycetae' (1922). An illustrated account of the British gasteroid fungi is long overdue. However, a comprehensive catalogue and thorough review of the literature was published by Palmer (1968), and identification keys, including some extralimital species, were presented by Demoulin & Marriott (1981). The latter were produced to facilitate species identification in preparation for a mapping scheme introduced by the British Mycological Society. These maps are incorporated in the present volume. Since 1981, at least two further species have been recorded from Britain, including the exotic phalloid *Clathrus archeri*, which has become well established in southern and south-east England. In 1993, the first European outdoor record of another exotic phalloid, *Aseroe rubra*, was obtained from Surrey (Spooner 1994).

Acknowledgements

The authors are grateful to the British Mycological Society and to Dr. B. Ing for permission to publish distribution maps from the gasteromycetes mapping scheme. Dr. J. Cooper is thanked for skilful help in producing the camera-ready maps. The following are kindly thanked for supplying photographs for use or consideration: A. W. Brand, F. D. Calonge, M. Christensen, V. Demoulin, G. Dickson, J. & M. Jeppson, C. Lange, P. D. Livermore, E. Mrazec, J. Nitare, A. Outen, J. T. Palmer, E. Rald, P. Roberts, M. Rotheroe, P. R. Smith, M. V. Strong, S. Sunhede, J. Vesterholt and W. Winterhoff. The following are thanked for supplying various types of information during the project: E. G. Gange, A. Henrici, B. Ing, H. Kreisel, J. Skinner.

FORM AND STRUCTURE

The gasteroid fungi represent the widest range of form and structure found within any group of comparable size, ranging from just visible to the naked eye, such as the species of *Mycocalia* Palmer, to large and complex forms, such as the Giant Puffball, *Calvatia gigantea* (Batsch: Fr.) Lloyd. The genera and families within the group are of mixed ancestry and consequently they are difficult to define. They share only two unifying characteristics, namely:

(i) the basidium is not involved in the discharge of the spore.

(ii) the basidia are enclosed within the fruitbody (or **gasterocarp**), at least in the early part of their development, with a few exceptions of very reduced structure. Thus they are defined as **angiocarpic**.

Generally, the fruitbodies of gasteroid fungi are formed by an outer protective coat, the **peridium**, and an inner fertile tissue, the **gleba**, in which are developed numerous basidia, producing basidiospores. As with all macromycetes, the fruitbody is constructed of hyphae. The fruitbodies may develop above ground level, at least in the later stages, and are described as epigeous, or they may be entirely subterranean in development, forming false truffles, and are described as **hypogeous**.

Basidium

The basidium of most gasteroid fungi is a **holobasidium**, which is neither septate nor segmented. This is a remarkable structure, for the single cell is responsible for nuclear fusion (karyogamy), meiosis, and spore production. Unlike the gymnocarpic macromycetes, the spores of gasteroid fungi are not actively discharged but released by passive detachment.

In a number of genera e.g. Sclerodermatales, the basidia are not orientated to form a palisadic hymenium but developed in small clusters within the gleba; these are called **plectobasidia**. It has been postulated (Dring 1973) that such plectobasidia are derived from a relatively simple basidiomycete ancestor in which the basidia were not genetically fixed but acted only in the role of 'spore guns', and were not marshalled into a hymenium. Several generations of plectobasidia may form within the fruitbody.

Basidia are very variable in the gasteroid fungi. Those with the simplest structure are to found in the false truffle genus, *Hymenogaster* Vittad., in which the basidium is narrow, and hypha-like, producing just two straight sterigmata (Fig. 1A). In *Podaxis* Desv., a tropical, stipitate puffball genus belonging to the *Agaricales*, the basidium is very inflated so as to appear subglobose (Fig.1B). Some genera have basidia in which the sterigmata are developed at the expense of the main body e.g. *Lycoperdales* (Fig. 1C), in which the long sterigmata may eventually break away but remain attached to the spore whilst, in other examples, the sterigmata are reduced resulting in an almost sessile spore e.g. *Astraeus* Morgan (Fig. 1D). The positioning of the sterigmata is not always apical; they can become laterally attached to the basidium, e.g. earthstars (*Geastraceae*), earthballs (*Sclerodermatales)* or stalked puffballs (*Tulostomatales)* (Fig. 1E). In addition, the number of sterigmata can vary in number, often being eight in, for example, earthstars and stinkhorns (*Phallales*) (Fig. 1F).

Basidiospores

Most gymnocarpic macromycetes are ballistosporic, the basidiospore being forcibly discharged from the tip of the sterigma. In the gasteroid fungi, however, the basidiospores

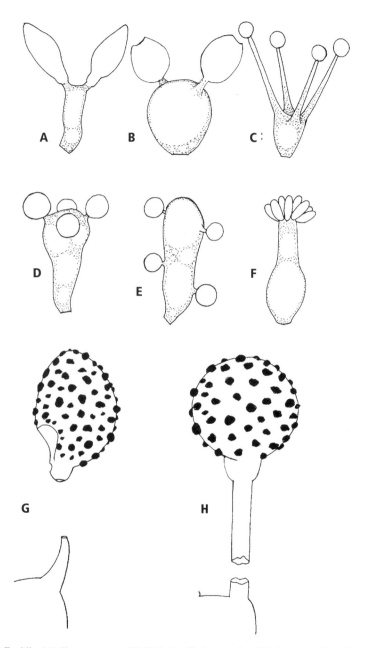

Fig. 1. (A-F) Basidia. (A) *Hymenogaster*; (B) *Podaxis*; (C) *Lycoperdon*; (D) *Astraeus;* (E) *Tulostoma*; (F) *Phallus*. (G) Ballistospore, typical of a gymnocarpic macromycete, showing heterotropy, bilateral symmetry with flattened adaxial and expanded abaxial surfaces, suprahilar depression, hilar appendix with punctum lacrymans, arising from a curved sterigma; (H) Statismospore, typical of an angiocarpic macromycete (gasteromycete) showing orthotropy, radial symmetry, no suprahilar depression, hilar appendix with attached sterigmal appendage (pedicel), arising from a straight sterigma.

exhibit **statismospory**, with no active discharge from the basidium but a passive falling off at or near maturity. **Ballistospory** (Fig. 1G) necessitates not only a specialised sterigma but a peculiar basidiospore structure and development. The spore primordium forms excentrically (heterotropic) on the apophysis of the sterigmal apex, and the spore itself is asymmetric, with bilateral symmetry resulting from the relative underdevelopment of the adaxial surface, a strongly expanded abaxial surface and usually a suprahilar depression. In addition the hilar appendix shows a subterminal hilar scar plus an adaxial *punctum lacrymans*, where the droplet necessary for ballistospory has been extruded. The **statismospore** (Fig. 1H) of the gasteroid fungi shows none of these features. The position of the spore initial on the sterigmal apex is orthotropic, and the subsequent expansion is one of radial symmetry. Instead of a hilar scar there is generally either a torn, open terminal hilum or the attachment of a persistent sterigmal appendage, termed the **pedicel** e.g. *Lycoperdon* Tourn.: Pers. The hilar appendix is cylindrical rather than tapering. Some gasteroid spores e.g. *Phallales*, the slime truffles (*Melanogastrales*), and the milk cap truffles (*Elasmomycetaceae*), do have a tapering hilar appendix with a hilar scar, possibly indicating an ancestral relationship with the *Agaricales*.

The basidiospores are highly characteristic for each genus, and therefore offer the best guide to phyletic relationships. In most cases the spore is pigmented and ornamented, whilst in others it is hyaline and smooth e.g. *Phallales*, the bird's nest fungi (*Nidulariales*).

The spores of the *Sclerodermatales* are shed at an immature stage and development continues with the help of a placenta of nurse cells (trophocysts). It has been suggested (Dring, 1973) that the trophocysts may be a second generation of modified basidia.

Hyphal structure

The fruitbodies are formed by hyphae which may either be all of one type or comprise specialized hyphal types. When only one hyphal type is present, the hyphal system is termed **monomitic**, and the hyphae are regularly septate, branching, and of unlimited length, with or, more commonly, without clamp-connexions. In some of the tougher fruitbodies, a second hyphal type, the **skeletal hypha**, is formed by the generative hypha, and this is unbranched, non-septate, thick-walled, often pigmented and of limited growth. Such a hyphal system is termed **dimitic**, and may be compared to the structure found in the more coriaceous gymnocarpic macromycetes.

A specialized form of skeletal hypha is the **capillitial thread**, which collectively comprise the **capillitium**. This is a tangled mass of hyphae, which are only to be found in forms with a powdery gleba. **True capillitial threads** are confined to the *Lycoperdales* and have thickened, brown pigmented walls, which may have pores e.g. *Calvatia* Fr., slits e.g. *Handkea* Kreisel or neither. They are generally branched but in some cases only very sparingly so. Capillitial threads are cyanophilic, positively staining with aniline-blue or cotton-blue. In *Bovista* Pers.: Pers. and *Bovistella* Morgan, the capillitial thread forms many short, tapering branches arising from a central stem. The branches are stellately arranged, about 350 - 400 µm long (Fig. 2A). Such a hypha resembles the skeleto-ligative hypha of the poroid macromycetes. In *Geastrum* Pers. and *Lycoperdon* Tourn.: Pers., the capillitial hyphae (Fig. 2B) show occasional branching only. In the stilt puffballs, *Battarraea* Pers., there are reduced capillitial threads called **elaters** which are free, short elements up to 150 µm long. These taper at the ends and have spiral thickenings (Fig. 2C). They are hygroscopic and, like all capillitial threads, assist dissemination of the powdery spore mass. In addition to the

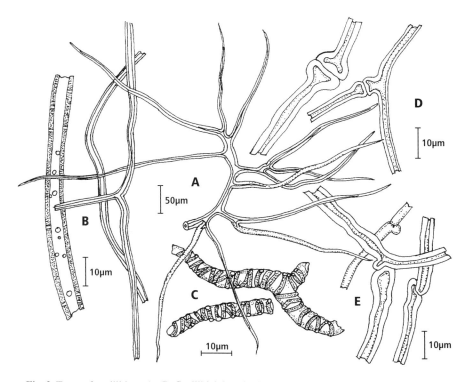

Fig. 2. Types of capillitium. A - B, Capillitial threads. A, *Bovista* - type; B, *Lycoperdon* - type. C, Elaters of *Battarraea*. D - E, Paracapillitial threads. D, *Tulostoma* - type; E, *Astraeus* - type.

elaters, thin-walled, hyaline hyphae are also present in the *Battarraea* gleba, and these are termed **pseudocapillitial hyphae**.

Sometimes, **paracapillitial threads** are present, which are modified generative hyphae with septa. They usually remain unpigmented e.g. *Tulostomatales* and *Astraeaceae*, and are acyanophilic. In *Tulostoma* Pers.: Pers., the paracapillitial hyphae are hyaline or pigmented with a distinctly thickened wall, and characteristically swollen at the septa (Fig. 2D), whilst in *Astraeus*, the septa have conspicuous clamp-connexions (Fig. 2E). In *Vascellum* Smarda, numerous thin-walled, septate paracapillitial hyphae are found in the gleba, especially towards the peridium, alongside the pigmented, true capillitial hyphae.

Structure of fruitbody

The fruitbody comprises two main tissues: a sterile, outer wall, termed the **peridium**, which surrounds and protects the fertile inner tissue, the **gleba**. The **peridium** may consist of one layer or be stratified, with up to three layers, termed **exoperidium, mesoperidium** and **endoperidium**. The **exoperidium**, or outer layer, is the first to break down and this can produce distinctive patterns over the surface, together with granules and spines, often characteristic for each species. The **mesoperidium**, when present, is often gelatinized, offering further protection to the developing basidia. The **endoperidium** usually forms a tough membrane, which is persistent and forms the 'spore-sac' in the *Geastraceae*.

In the early stages of development the gasterocarp primordium forms an egg-like body, the **mycoegg**, which is often hypogeous and requires protection by the tough peridium. In the false truffle genera, the peridium generally remains entire and closed. In the forms which develop epigeous fruitbodies at maturity, the peridium may persist as a basal volva e.g. *Phallales*. In the *Lycoperdales*, the peridium stops growing at maturity, and progressively breaks down, mainly through weathering. In the *Nidulariales*, the peridium serves as specialised cup to catch rain-drops, enabling spore dispersal by splashing.

The **gleba** comprises the spore mass plus any associated structures. In the mature gasterocarp the gleba may be powdery e.g. in *Geastrales*, *Lycoperdales*, *Sclerodermatales* and *Tulostomatales*, mucilaginous by autolysis e.g. in *Melanogastrales* and *Phallales*, or form spore packets (**peridioles**) e.g. in *Nidulariales*. The tissue of the gleba may be continuous or form isolated or labyrinthoid **locules**, in which the basidia are developed. The gleba may develop in one of six ways, which have been summarised by Kreisel (1980) and Miller & Miller (1988). The glebal types have been used as the most fundamental criterion for the classification of gasteromycetes (Dring,1973), and may be briefly summarised as follows:

(i) **Homogeneous**: basidia evenly distributed throughout the gleba e.g. *Tulostomatales* (Fig. 2A).
(ii) **Lacunate**: gleba develops regular chambers, each lined with a hymenium e.g. *Nidulariales*, *Melanogastrales* (Fig. 2B).
(iii) **Forate**: groups of plectobasidia are initiated towards the periphery of the gleba and progressively extend inwards (centripetally) e.g. *Lycoperdales* (Fig. 2C).
(iv) **Aulaeate**: plectobasidia formed on down-growing primordial tissue from the upper gleba, the tissue then fragments to form locules e.g. *Hymenogastrales*; *Gastrosporium* (Fig.2D).
(v) **Unipileate**: a central stalk (receptacle) giving rise to an apical gleba e.g. *Phallaceae* (Fig. 2E).
(vi) **Multipileate**: a series of secondary stalks (two or more), each developing its own gleba e.g. *Clathraceae* (Fig. 2F).

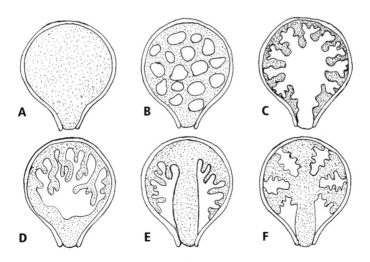

Fig. 3. Development of the Gleba. (A) Homogeneous; (B) lacunate; (C) forate; (D) aulaeate; (E) unipileate; (F) multipileate.

Gasterocarp form

The form and overall structure of the gasterocarp is directly related to the spore dissemination mechanism. Gasterocarps may be sessile or stipitate. In the forms with a stem, there may be a **true stipe** e.g. *Tulostomatales*, constructed of longitudinally orientated hyphae, or a **pseudostipe** e.g. *Lycoperdales*, in which the stem consists of a spongy tissue without longitudinal arrangement, or a **receptacle** e.g. *Phallales*, a specialised structure with a spongy texture which rapidly expands to lift the gleba above ground level.

Spore dispersal may be by one of three means, resulting from either a powdery gleba, a mucilaginous gleba, or by the formation of spore packets. Typical puffballs form a powdery gleba, which disintegrates at maturity, resulting in enormous numbers of spores which are released by rain-splash as a cloud into the air-currents. In *Astraeus*, *Geastrales* (Fig. 3B), *Lycoperdon* (Fig. 3A), and *Tulostomatales* the peridial apex develops an opening, the **peristome**, through which spore-clouds are ejected following the splash action of rain-drops, or other pressure, on the peridial surface. The genus *Myriostoma* Desv. develops several peristomes over the peridial apex, suggesting the common name of 'Pepper Pot'. In the Giant Puffball (*Calvatia gigantea*), the peridial layers flake away and the gleba progressively breaks down releasing spores over an extended period. In *Bovista*, the entire gasterocarp becomes detached and is blown about by the wind, eventually releasing its spores by the peridium breaking open. Members of the *Sclerodermatales* (Fig. 3E) similarly expose the powdery gleba by irregular fracture of the peridial wall.

In the *Phallales* and *Melanogastrales* an entomophilous method has evolved for spore dispersal. At maturity the gleba autolyses into a slime with a characteristic odour, and this is attractive to insects, mainly Diptera. The slime, containing spores in great numbers, adheres to the insect body and the spores are thereby disseminated. The *Phallales* demonstrate a considerable range of additional structures to raise the gleba above ground level. In all cases a **receptacle** is formed, this may be a cylindrical stem-like structure, as in *Mutinus* Fr. and *Phallus* Hadr. Jun.: Pers. (Fig. 3F), or the receptacle may form either a lattice-structure or stellate rays, as exemplified in the *Clathraceae*.

Rather than releasing individual spores, the *Nidulariales* develop spore packets. The Bird's Nest Fungi (*Nidulariaceae*) (Fig. 3C), form lenticular spore-packets, called **peridioles**. The peridiole consists of a packet of glebal tissue, surrounded by a tramal plate which forms a protective wall; the basidia within the peridiole produce large numbers of spores. The peridium develops as a cup-like structure which contains one to many peridioles, and the latter are released as a unit by a rain-drop splash action in the cup. The tiny Cannon Fungus (*Sphaerobolus*) (Fig. 3D) has developed an active mechanism for discharging its single peridiole.The inner peridial layer violently everts to catapult the peridiole a distance of up to four metres.

Finally, mention should be made of the secotioid fungi, although they are not found in the British Isles. These are fungi which are closely related to the *Agaricales, Boletales* and *Cortinariales*, and closely resemble them in form. It is likely that they have evolved from ballistosporic, agaricoid fungi by reduction. The peridium forms a pileus, which is centrally stipitate, and the hymenophore resembles radial lamellae although modified by interveining and branching to produce a labyrinthoid gleba. The pileus may or may not open to expose the gleba, and the spores have lost the ability of active release and are, therefore, statistosporic although resembling a ballistospore in structure.

Gasterocarp form

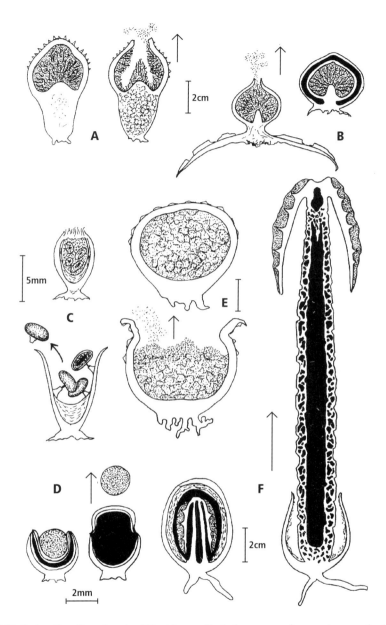

Fig. 4. Vertical sections through gasteroid basidiomes, illustrating a range of spore release mechanisms. A, *Lycoperdon*, a puffball, with a single-layered peridium opening at maturity by an apical ostiole to release powdery spores. B, *Geastrum*, an earthstar, the outer peridial layers split radially, curving back to lift the spore-sac above ground level. C, *Cyathus*, bird's nest fungus, the spores are developed within hard packets (peridioles), which are ejected by rain splash-action. D, *Sphaerobolus*, cannon fungus, the inner peridial layers invert suddenly to violently eject the spore-mass. E, *Scleroderma*, an earthball, the thick, single-layered peridium breaks open irregularly to release powdery spores. F, *Phallus*, a stinkhorn, the gelatinous myco-egg splits and the slimy spore-mass is raised above ground-level by a rapidly elongating receptacle.

DISTRIBUTION AND HABITATS

Many of the species covered in this book have a very wide distribution, although most tend to be restricted to particular climatic zones. Temperate zone species can occur at high altitude in the tropics. Epigeous fungi have adapted to a very wide range of habitats but probably they exhibit greatest species diversity in the drier parts of the world. The stinkhorns and allied fungi, however, are mostly found in the humid tropics.

The British Isles are sufficiently diverse in terms of climate and geology to offer a wide range of suitable habitats. Nevertheless, several species on the European mainland have not been found in Britain and possibly await discovery. A few species have almost certainly been introduced through human activity. This is true for some stinkhorn relatives, e.g. Devil's Fingers, *Clathrus archeri* (Berk.) Dring. Similarly, a few of the earthstars, especially those favouring disturbed sites and often associated with exotic trees, may also have been introduced.

Several species are no longer found in British Isles. Ing (1992) listed eight species in the 'extinct category' for his provisional 'Red List of Endangered Species'. One example is the Pepper Pot, *Myriostoma coliforme* (With.: Pers.) Corda, which, although originally described from Britain and recorded from nine English localities, has not been seen for more than a century. This species was first collected in Kent by John Ray in 1695 'in a lane leading from Crayford to Bexley-Common' (Ray 1724: 28) and again reported by Ray from Hampton Court Palace. The last known collection was in 1880, when specimens were found at Hillington, Norfolk 'in a hedge bank in a green lane in a large nettle clump' (Plowright, 1881). A number of other species have only been recognized and recorded in recent years. For example, *Tulostoma niveum* Kers was discovered in a remote Scottish site in 1992, yet had hitherto been known only in Sweden. *Aseroe rubra* Labill. is probably a very recent introduction, having been discovered in Surrey (Spooner, 1994). This species is Australasian in origin and probably entered Britain through human activity.

There are no endemic species amongst those described in this book. Nevertheless, some of the less common should be considered for conservation attention. This mainly means preserving and maintaining proper management of the habitat. One of the few known sites for *Battarrea phalloides*, a hedgerow, is managed to ensure the survival of this rare species. Ing (1992) listed a further seventeen species in other categories, such as 'endangered' or 'vulnerable'.

All the species are listed in Table 1, indicating their preferred habitats. The table shows that many species may occur in only one, or a few, of the broadly defined habitats. Some species can be seen to occur exclusively in dunes, e.g. *Phallus hadrianus* Vent.: Pers. and *Cyathus stercoreus* (Schwein.) De Toni, whilst others occur in very wet sites, e.g. most *Mycocalia* species. Many species thrive in types of grassland, whilst others are confined to woodlands. Most species are saprophytes, developing in the soil and humus, but some are able to live on dead wood substrates, e.g. *Lycoperdon pyriforme* Schaeff.: Pers., at least when in the later stages of decay. The earthballs, *Scleroderma* spp., apparently form mycorrhizal associations with a wide range of trees. Dyeball, *Pisolithus arhizus*, is a proven mycorrhizal species, forming associations with a wide range of trees, though apparently only with *Pinus* in Britain. Its mycorrhizal capacity is used to facilitate reclamation of polluted soil, such as that found at mine spoils, by infection of seedlings for planting. Certain species, such as some earthstars and the Giant Puffball form fairy rings. The spreading of wood-chippings onto ornamental beds has greatly increased habitat availability to some of the bird's nest fungi, and members of the stinkhorn order, e.g. *Clathrus archeri*, are also flourishing in this new habitat. Dung is an unusual habitat but two

Distribution and Habitats

species can regularly be found on it. *Cyathus stercoreus* occurs only on rabbit pellets amongst marram grass dunes.

Information on the ecological requirements of grassland associated species was reviewed by Arnolds (1981, 1982). A more general treatment may be found in Schmitt (1978) and Demoulin (1969).

TABLE 1. Habitat preference for species covered

A: gardens, parks, flower beds with wood chippings, road verges, cultivated fields, etc. **B**: calcareous woodland, typically with beech (broad leaved). **C**: calcareous grassland. **D**: dunes. **E**: fens and bogs. **F**: grassland, neutral to acid. **G**: heath. **H**: improved grassland. **J**: neutral to acid woodland, typically with oak and birch. **K**: conifer plantations and native pine woods.

	A	B	C	D	E	F	G	H	J	K
Scleroderma areolatum									J	
Scleroderma bovista	A								J	
Scleroderma cepa	A								J	
Scleroderma citrinum							G			
Scleroderma polyrhizum									J	
Scleroderma verrucosum	A	B					G		J	
Pisolithus arhizus	A									K
Astraeus hygrometricus	A								J	K
Tulostoma brumale			C	D						
Tulostoma melanocyclum				D						
Tulostoma niveum			C							
Queletia mirabilis	A									
Battarraea phalloides	A									
Cyathus olla	A			D						
Cyathus stercoreus				D						
Cyathus striatus		B							J	
Crucibulum laeve	A	B							J	
Nidularia deformis	A									K
Mycocalia denudata					E					
Mycocalia duriaeana					E					
Mycocalia minutissima					E					
Mycocalia sphagneti					E					
Sphaerobolus stellatus	A		C	D	E	F		H		
Geastrum berkeleyi		B	C							K
Geastrum campestre	A		C							
Geastrum corollinum	A	B								
Geastrum coronatum	A	B	C							
Geastrum elegans	A			D			G			K
Geastrum fimbriatum	A	B								K
Geastrum floriforme	A									
Geastrum fornicatum	A	B								
Geastrum lageniforme	A									

TABLE 1 cont. Habitat preference for species covered

A: gardens, parks, flower beds with wood chippings, road verges, cultivated fields, etc. **B**: calcareous woodland, typically with beech (broad leaved). **C**: calcareous grassland. **D**: dunes. **E**: fens and bogs. **F**: grassland, neutral to acid. **G**: heath. **H**: improved grassland. **J**: neutral to acid woodland, typically with oak and birch. **K**: conifer plantations and native pine woods.

	A	B	C	D	E	F	G	H	J	K
Geastrum mininum				D						
Geastrum pectinatum	A									K
Geastrum quadrifidum		B								
Geastrum rufescens		B		D			G			
Geastrum schmidelii				D						
Geastrum striatum	A	B								K
Geastrum triplex	A	B								
Myriostoma coliforme	A									
Bovista aestivalis		B		D			G			
Bovista dermoxantha	A					F				
Bovista limosa			C	D						
Bovista nigrescens	A					F	G	H		
Bovista paludosa					E					
Bovista plumbea	A		C	D		F		H		
Bovistella radicata	A					F				
Handkea excipuliformis	A	B	C			F			J	K
Handkea utriformis			C	D		F	G			
Calvatia gigantea	A		C					H		
Lycoperdon atropurpureum									J	
Lycoperdon caudatum				D						
Lycoperdon decipiens			C							
Lycoperdon echinatum		B								
Lycoperdon ericaeum						F	G			
Lycoperdon lambinonii										K
Lycoperdon lividum			C	D						
Lycoperdon mammiforme		B								
Lycoperdon molle		B	C							
Lycoperdon nigrescens	A			D		F	G		J	K
Lycoperdon perlatum	A	B							J	K
Lycoperdon pyriforme	A	B							J	
Lycoperdon umbrinum										K
Vascellum pratense	A			D		F		H		
Phallus impudicus	A	B		D					J	K
Phallus hadrianus				D						
Mutinus caninus		B							J	
Mutinus ravenelii	A									
Clathrus archeri	A	B								
Clathrus ruber	A	B								
Aseroe rubra									J	
Ileodictyon cibarium	A									
Lysurus cruciatus	A									

HOW TO COLLECT AND EXAMINE GASTEROID FUNGI

Gasteroid fungi can be found in a wide range of habitats, and it is important therefore to understand the habitat requirements of the different species. These are outlined in the previous chapter, and more fully discussed under each species. These fungi also exhibit a great range in size, and the tiny *Mycocalia* species, for example, require specialist attention; careful searching in wet places at the stem bases of rushes and sedges, and amongst mosses is required to find the fruitbodies. Most species fruit during summer and autumn, but many of the bird's nest fungi may occur at other times, and some phalloids, notably Clathraceae, can occur in winter. Fruitbodies of earthballs and some puffballs are tough and do not readily break down, so that they can be found at all times of the year. However, old specimens are usually of little value to a collection as they are likely to have lost many of their important characters, although earthstars are often easy to name from overwintered fruitbodies. It is important that all stages of the fungus be obtained if possible. Mature fruitbodies may be collected without other stages if necessary, but young specimens alone should be avoided. If necessary, return to the site through the season to obtain the different developmental stages. Many gasteroid fungi can be messy to collect; earthballs and puffballs have powdery spore masses which easily contaminate other collections, and phalloids have a slimy, unpleasant-smelling gleba. If collected, these should be placed separately in a box or tin with a tight fitting lid. Puffballs can also be wrapped separately in waxed paper. It should be noted that phalloids can be photographed in many cases, and for these it may be unnecessary to take specimens.

All gasteroid fungi except phalloids are tough and can be left to air dry. They can then be stored in packets or boxes which must be clearly labelled with full details of the collection: species name, habitat, locality, date, collector and any relevant descriptive notes. Phalloids putrify readily and should be dried rapidly, preferably not in the house! They are generally quite fragile after drying and need careful storage to prevent damage. They can also be preserved in 70% alcohol, a method which cannot be recommended for any other group of fungi. This will shrink them and leach the colours but preserve their overall form, a character which is important and can be lost in poorly dried material. It should be noted that phalloids are scarcely ideal subjects for private herbaria. Dried specimens should be deep frozen for at least 48 hours to rid them of insects and mites. It may be noted that puffballs, earthstars and earthballs all support a specialised insect fauna comprised especially of beetles and flies. In Britain, at least 7 species of beetle from 4 families (*Anobiidae, Cryptophagidae, Endomychidae* and *Nitidulidae*) are associated with these fungi, completing their whole life cycle within the fruitbodies. Flies from at least 7 families are also associated with these fungi and others occur with *Phallus* and with other members of the *Phallales*.

Identification of many gasteroid fungi requires the use of light microscopy for examination of spores, capillitial and paracapillitial characters. For puffballs and others having a powdery spore mass, a small portion of the gleba should be removed using a needle or fine forceps, and mounted initially on a glass slide in a drop of distilled water or 5% ammonia solution. Use of alcohol (ethanol) for initial wetting of the spore mass before mounting may be useful in some cases. Tap or tease out and mount under a coverslip. Staining is often unnecessary for pigmented hyphae, but use of cotton blue in lactophenol may facilitate examination of ornamented spores, especially those of *Scleroderma* species, and Melzer's Reagent is acceptable for spores of *Lycoperdon* and *Geastrum* species. For examination of basidia and thin walled hyphae, staining with a dilute solution of phloxine or Congo Red in 5% KOH is helpful. The gleba of phalloids and related fungi may be examined in distilled water.

FOLKLORE AND USES

The fungi considered in this book offer a wide range of unusual forms which have contributed to the development, in many parts of the world, of a rich and interesting folklore. Much of this stems from times when fungal fruitbodies were unexplained developments which inevitably received magical connotations. However, a range of practical uses, including value as food and in medicine, have also been found for many gasteroid species. Some of these aspects are considered here, with emphasis on those applying to the British Isles.

The folklore of gasteroid fungi mainly concerns puffballs and stinkhorns, which have influenced cultures worldwide, often in similar ways. Puffballs are nowadays mainly regarded as a curiosity, either because of their unusual method of spore dispersal - they puff when squeezed - or, in the case of the Giant Puffball, because of its impressive size. However, they figure prominently in the folklore of many cultures and also had religious or magical significance to some North American Indian tribes. The Blackfoot, for example, referred to them as 'fallen stars' and used them as incense to keep away ghosts. This and other examples are summarized by Burk (1983).

There are many folk names applied to puffballs, some recorded in the British Isles as early as the 16th century, such as 'bulfists', 'pulker-fist', 'pucke fusse' and 'molly-puffs' (Gerard 1597; James 1747). Others include 'fusse bals', 'puff fistes', and 'bofist'. Such names are said to reveal an association with the supernatural, referring either to the Devil (old English 'pouke') or Puck, the fairy Robin Goodfellow (e.g. Steele 1888; Rolfe & Rolfe 1925). They are clearly reflected in the generic name *Bovista*. However, the etymology of these names is complex and uncertain. Another possible derivation is from 'pogge', meaning toad, hence also the word 'toadstool' (Folkard 1884; Friend 1886), although this appears at least as early as the 14th century (Rolfe & Rolfe 1925). Wasson & Wasson (1957) link their origin with breaking wind rather than with the supernatural, and it is interesting to note that the generic name '*Lycoperdon*' translates as 'wolf breaking wind'. Folk names with a similar derivation occur elsewhere in Europe. In Denmark, for example, they include 'ulvefis', which was in use by the 17th century and has the same meaning as the name *Lycoperdon* (Brøndegaard 1987).

Various other folk names reflect the powdery spore mass of mature fruitbodies. Amongst these are 'devil's snuff-box', 'old man's snuff-box', 'pixie-puff', 'blindman's ball' or 'bellows' and, in Scotland, 'blind men's een'. The last two refer to the common belief that the spores are damaging to the eyes. A similar belief was held by some North American Indian tribes who called puffballs names such as 'no-eyes' or 'ghost's makeup' (Burk 1983). The terms 'poor-blinde' and 'sand-blinde' were used by Gerard & Johnson (1633) to describe the effects of the spores of puffballs in the eyes, and this is also reflected in the Danish name 'blinderøg'. According to Wasson & Wasson (1957), there are more folk names for puffballs than for any other fungi.

Earthstars, surprisingly perhaps in view of their unique form, appear to have little associated folklore. However, in the 17th century, a human-like form was seen in *Geastrum fornicatum* by Seger, who referred to it as 'Fungus anthropomorpha', depicting a ring of human-like figures.

The folklore of stinkhorns, as might be expected, largely concerns their smell and phallic shape. However, they have also been associated with death and the devil as, for example, in parts of Europe, America and Borneo (Brøndegaard 1983). In Borneo, the stinkhorn symbolised the penis of a dead hero, now returning in spirit form. In Germany, stinkhorns appearing in a graveyard were regarded as the fingers of a corpse pushing up from the grave, indicating unrepented sin in the dead (Brøndegaard 1983). They were known as 'Leichenfinger' ('corpse finger') or 'Totenpilz'. In Massachusetts, the stinkhorn was known as a 'death baby', and its occurrence near the house was considered as a sign of imminent death in the family (Brøndegaard 1983). In Europe, the eggs of the stinkhorn were once thought to be those of evil spirits or devils (Findlay 1982). Names such as 'ghost's egg' and 'devil's egg' applied to them in England; in Sweden they were known as 'trolläg', and in Germany as 'Hexeneier' (Brøndegaard 1983).

The name 'stinkhorn' has long been in use, and popular names for this fungus are many, sometimes reflecting its smell, as for example 'devil's stinkpot' (Yorkshire), 'pow-cat', 'carrion flower' (Brøndegaard 1983) and 'stinking polecat fungus' (Miller 1884, Smith 1882). Others are direct references to its phallic shape: 'pricke mushroom' (Gerard 1597), 'Devil's horn' (Miller 1884, Steele 1888), and 'Satan's member' (Emboden 1974). In pre-Linnean literature, before the introduction of binomials, it was even known as 'fungus foetidus penis imaginem referens' and 'fungus virilis penis arrecti facie'. Other names are given in Wasson & Wasson (1957). The provocative shape of the stinkhorn inevitably led, on the one hand, to its use as an aphrodisiac, and on the other to its causing of severe offence in puritanical circles. Its supposed aphrodisiac properties led to its sale in parts of Europe, especially in Germany, during the Middle Ages. However, there is no physiological evidence in support of its effect as an aphrodisiac.

The offence taken at the form of the stinkhorn is exemplified by the attitude of Darwin's eldest daughter Etty. According to her niece Gwen Raverat in her book of Victorian reminiscences 'Period Piece', published in 1953, she would scour the local woods for stinkhorns and then bring them home to burn them 'behind closed doors'. The story is recounted more fully in Findlay (1982) and in Wasson & Wasson (1957). A similar attitude was held by Beatrix Potter who 'could not find courage to draw it' (Jay et al. 1992). One of the earliest illustrations of the species was by Clusius (1601), whose figure was later reproduced, but tellingly upside down, by Gerard & Johnson (1633) in that edition of the famous herbal. This theme is further explored by Wasson & Wasson (1957).

Aphrodisiac properties were also attributed to some of the bird's nest fungi. In Colombia and Guadeloupe these were reputedly used to stimulate fertility (Dickinson & Lucas 1979; Brodie 1975). There seems to be less associated folklore with these fungi, although they have received some folk names, such as 'elfin cups', 'fairy goblets', and 'corn bells' (Brodie 1975). In England they were also known as 'pixies' purses' and in Scotland as 'siller cups' (Steele 1888). To find them on the way to work was once regarded by Scottish country folk to be a lucky omen for the day (Steele 1888). These fungi certainly came to the attention of early naturalists, although their nature was misunderstood. The 'eggs' or peridioles were, not unreasonably, considered to be seeds, although this led to the interesting claim by Goedart that when 'cherished by the sun' these sprouted feet and began to live as birds. This he had observed, with careful scrutiny, over two successive years! (Emboden 1974).

Puffballs, especially the Giant Puffball, have long had culinary value. They were evidently well known to the Greeks and Romans (Buller 1915), and their use in this respect remains widespread. Most species are edible in their young stages before the spores develop. However, such use is not universal. The North American Iroquois, for example, apparently referred to puffballs as 'Devil's bread', although amongst other Indian tribes they were evidently an important food item (Burk 1983). According to Smith (1951), some species of *Calvatia* and *Lycoperdon* have been implicated in 'violent gastrointestinal upsets', though this may result from eating specimens which are nearing maturity. Earthballs (*Scleroderma* spp.), in contrast, are known to contain toxins (Lincoff & Mitchell 1977), and they should never be consumed in more than the tiniest quantities. Nevertheless, they have been used as adulteration for truffles and as a condiment.

The stinkhorns and their allies have little culinary value, although the unexpanded eggs and the spongy stipes are edible and have been used in salads. In places, the eggs are pickled, and in Germany they are incorporated into sausages and are also sold as a kind of truffle (Dickinson & Lucas 1979). In China, *Lysurus mokusin* is eaten (Berkeley 1857), and is said to be a great delicacy (Dickinson & Lucas 1979). The gelatinous volva of species of *Ileodictyon* is eaten in New Zealand (Berkeley 1857), where the fungus is known by the Maoris as 'thunder dirt' (Steele 1888). *Clathrus ruber* is also said to be edible in the egg stage (Marchand 1976), although Barla (1859) describes a case of poisoning by this species. Cooke (1862) also gives an alarming report of the development of violent convulsions and loss of speech in a young girl after eating this fungus. In France, it has, quite without foundation, been feared as a cause of cancer, skin lesions, sickness, convulsions and delirium (Dring 1980; Dickinson & Lucas 1979). It is referred to there as 'cancrou' or cancer (Badham 1863).

Medical uses for puffballs, stinkhorns, and other gasteroid fungi are many and widespread. Direct use of the fruitbodies has a long history, but drugs and other active components have been isolated from some species.

The most frequent use of puffballs is probably as a wound dressing. The Giant Puffball has been used to form a surgical dressing (Rolfe & Rolfe 1925), and the powdery spore mass of many species is a useful and effective styptic. Such use was widespread in Europe and North America, and was also found in India (Rai et al. 1993). According to Swanton (1917), larger puffballs were at one time commonly kept in farmhouses and cottages in West Sussex, and the custom still lingered on in the season of 1916. It should be noted that the inhalation of the spores of puffballs can cause a lung disease known as lycoperdonosis (Henriksen 1976).

In China, puffballs and related fungi of many genera are considered to be medically important (Ying et al. 1987). They are mostly used as a styptic, but many are also considered to reduce swelling and to detoxify. Some species are used for throat ailments, and others to reduce fever and coughing, or as a painkiller. In Himalayan areas such as eastern Nepal, use of puffballs as a cure for sprains was evidently practised by the Lepchás, smouldering fruitbodies being applied directly to the skin (Hooker 1854) .

The use of *Tulostoma* by the North American Ramah Navaho to cure leg bone fractures in sheep, by application either as a poultice or as an infusion, was given by Burk (1983), as well as the use by the Paiute of *Battarraea* and other puffballs for swellings and sores. In Namibia, the spores of a species of *Battarraea*, mixed with oils to form a protective skin ointment and a cosmetic, are used by the Topnaar people (Van den Eynden et al. 1992). The spores of earthstars were used by some North American Indians to cure discharges from the ear, and puffball spores were also used as a remedy for earache (Burk 1983).

Other uses to which Giant Puffballs have been put include treatment of diarrhoea in calves in Finland (Dickinson & Lucas 1979) and Germany (Brøndegaard 1987), and a similar and current use for *Bovista graveolens* in southern Bohemia was reported by Kotlaba (1955). In north America, the Potawatomi Indians use *Morganella subincarnata* as a cure for headaches and refer to it as the 'headache berry' (Smith 1933; Burk 1983). Some puffballs, including several Mexican species of *Lycoperdon* and *Vascellum*, have also been reported to have hallucinogenic properties, although they appear to have no active principles (Ott et al. 1975).

The anti-tumour drug 'Calvacin' has been extracted from the Giant Puffball (Beneke 1963) and other species, notably *Bovistella radicata*, *B. sinensis*, *Pisolithus arhizus* and *Scleroderma cepa* (Ying et al. 1987). This compound is present only in young fruitbodies and in very tiny quantities, but higher yields are reported from cultures. More recently, calvatic acid, another active factor with antibiotic and anti-tumour properties, has been isolated from another species, *Calvatia craniiformis* (Umezawa et al. 1975). Another drug complex, Cyathin, has been extracted from the bird's nest fungus *Cyathus striatus*, and others. This complex includes seven different compounds (Brodie 1975) and is a fungal antibiotic and bacteriostatic. Some tribes from Madhya Pradesh, India, use species of *Cyathus* to soothe sore eyes (Rai et al. 1993), but there appear to be no other known medical uses for bird's nest fungi.

Stinkhorns appear to have had little medical significance in Britain, although in 1865 they figured prominently in discussions in The Times newspaper as a possible cause of cholera and other epidemics (Brøndegaard 1983). However, elsewhere in Europe and in China, stinkhorns have been used in a variety of ways, and they also featured in Greek herbalism (Wasson et al. 1978). In the Middle Ages, especially, they were used to treat epilepsy, rabies, kidney problems, gout, rheumatism, and bleedings. In China, species of *Dictyophora* are used as a treatment for dysentery and athlete's foot, and *Lysurus mokusin* is reputed to be a cure for gangrenous ulcers (MacMillan 1861, Dickinson & Lucas 1979). According to Ying et al. (1987) it has anticancer properties. *Phallus rubicundus* is also used in China to treat 'sores, subcutaneous ulcers, scabies, carbuncle and fistula' (Ying et al 1987), and is reputedly used in Central India against typhoid, and to ease labour pains (Rai et al. 1993). In Nigeria, *Phallus aurantiacus* and other stinkhorns are used by traditional Yoruba doctors as part of a preparation to cure leprosy (Oso 1976).

Puffballs have been put to a variety of practical uses other than the medical, culinary and religious aspects discussed above. It was suggested by Watling (1975) and by Watling & Seaward (1976) that the presence of *Bovista nigrescens* in prehistoric dwellings in Orkney, Scotland,and other sites in England could indicate their use as insulation against draughts in buildings. The latter authors also speculated on the use of puffballs in early times as kindling for fires. This use as tinder was given for *Calvatia gigantea* by Cooke (1862), who also described the production of amadou by addition of saltpetre to dried fruitbodies of *Handkea utriformis*.

One of the main uses for puffballs, especially *C. gigantea*, was in beekeeping. Fumes from the smouldering fruitbody, placed beneath the hive, calm the bees. Puffballs have been so used in Europe, including the British Isles (Swanton 1917, Dickinson & Lucas 1979), and throughout North America (Burk 1983), and it is a long established practice. The tranquilizing properties of the fumes are actually due to an excess of carbon dioxide rather than to any special anaesthetic properties. However, puffballs were once employed more generally as an anaesthetic, being successfully used in operations according to Berkeley (1860). MacMillan (1861) states that they deprive 'the patient of speech, motion and sensibility to pain, while he is still conscious of everything that happens around him'.

The North American Indians had a variety of other uses for puffballs, summarised by Burk (1983). These included use as a dusting powder or baby talc, as a rattle by medicine men, as necklaces, 'prized because of the delicate odor they gave off', and for children's games. The flesh of some species was used as a pin-cushion in parts of Denmark (Brøndegaard 1987), and the hygroscopic nature of *Astraeus hygrometricus* led to the use of this species as a hygrometer according to Rolfe & Rolfe (1925).

Pisolithus arhizus, popularly known as Dyeball due to the abundant yellow and purple dyes which it contains (Stevens & Kidd 1953), was used in various areas for dying of cotton, wool and silk.

There are few practical uses to which stinkhorns have been put. However, in China, the boiled liquid from species of *Phallus* and *Dictyophora* is used as a short-term food preservative (Ying et al. 1987). It is also interesting to note that stinkhorns, although they are not luminescent, emit, according to Dickinson & Lucas (1979), sufficient radiation to affect photographic film even through cardboard!

Fig. 5. *Bovista nigrescens*. From Bolton (1789, pl.118): 'may, while young, be eaten with safety; and has a taste much like that of the common mushroom. The powder is used here, for stenching blood, in small new wounds'.

KEYS TO ORDERS AND FAMILIES

1. Fruitbody small, 1–10 mm diam., globose or cupulate, sessile, often gregarious; hymenium absent, basidia borne singly or in groups; gleba forming one to many separate peridioles; spores often large, smooth, hyaline
.... **Bird's Nest Fungi & Cannon Fungi** (Nidulariales, p.50), 2

1. Fruitbody large and differently constructed; hymenium present or absent; peridioles absent or if present formed with a common wall; spores less than 10 µm long 3

    2. Fruitbody up to 3 mm diam., on wood or dung; peridium with stellate, apical dehiscence to reveal single, globose peridiole; inner peridiole wall everts to allow violent release of peridiole .... **Cannon Fungi** (Sphaerobolaceae, p.68)

    2. Fruitbody 1–10 mm diam., obconical, dehiscing either irregularly or by a circumscissile epiphragm; usually several, ellipsoid peridioles present.
    .... **Bird's Nest Fungi** (Nidulariaceae, p.50)

3. Development pileate or multipileate,with mucilaginous or soft-fleshy, deliquescent gleba borne on a rapidly expanding pseudostipe, or receptacle arising within the peridium; hymenium present; spores small, ellipsoid to cylindrical, hyaline to pale green, smooth .... **Stinkhorns & Cage Fungi** (Phallales, p.164), 4

3. Development not pileate, with mature gleba pulverulent, non-deliquescent and retained within the peridium; spores more or less globose, brown and ornamented 5

    4. Fruitbody unipileate, with the gleba borne externally on upper part of an unbranched, cylindrical, hollow receptacle .... **Stinkhorns** (Phallaceae, p.164)

    4. Fruitbody multipileate, with the gleba borne on inner surfaces of a reticulate to stellate receptacle .... **Cage Fungi** (Clathraceae, p.174)

5. Fruitbody truly stipitate, subtending the fertile body; hymenium absent, basidia in scattered groups throughout the gleba; capillitium present
.... **Stalk & Stilt Puffballs** (Tulostomatales, p.38), 6

5. Fruitbody sessile or forming a pseudostipe, lacking a differentiated stipe 7

    6. Peridial dehiscence either by a pore or irregular lobes; capillitial threads smooth
    .... **Stalk Puffballs** (Tulostomataceae), p.38

    6. Peridial dehiscence by circumscissile splitting; capillitial threads with spiral or annulate ornament .... **Stilt Puffball** (Battarraeaceae), p.48

7. Hymenium present, basidia maturing simultaneously; capillitium present; gleba at first whitish; spore ornament obtusely verrucose
.... **Puffballs & Earthstars** (Lycoperdales, p.70), 8

7. Hymenium not developed, basidia borne singly or in groups, maturing at different times; capillitium absent; gleba at first violaceous to purplish; spore ornament spinose to reticulate .... **Earthballs & Dyeball** (Sclerodermatales, p.19), 9

8. Outer peridium splitting radially to form stellate rays, revealing an endoperidium which is apically dehiscent; capillitium non-septate, typically unbranched and forming a pseudocolumella .... **Earthstars** (Geastraceae, p.71)

8. Peridium not splitting radially, dehiscence either by an apical pore or by apical fragmentation and flaking away; capillitium septate or not, typically branched, forming neither a pseudocolumella nor a columella
.... **Puffballs** (Lycoperdaceae, p.108)

9. Peridium 1-layered or not stratified; fracturing irregularly to expose the gleba; fruitbody globose to tuberous, sometimes with pseudostipe
.... **Earthballs** (Sclerodermataceae, p.20)

9. Peridium complex and stratified; splitting stellately from the apex; fruitbody geastroid, sessile .... **Barometer Earthstar** (Astraeaceae, p.36)

EARTHBALLS, DYEBALL and BAROMETER EARTHSTAR

SCLERODERMATALES G. Cunn.
Gasterom. Austr. & New Zeal. : 112 (1944).

Fruitbody more or less globose, tuber-like or 'geastroid', sessile or sometimes attached by a stem-like base (pseudostipe). *Peridium* often thick, dehiscing by irregular fracture at the apex or, rarely, by stellate splitting of the outer layer; 1–3-layered. *Gleba* lacunate in development, with tramal plates, which form 'peridiole'-walls in *Pisolithus*, pulverulent at maturity, lacking a capillitium. *Hymenium* not developed; basidia arising singly or in small fascicles in gleba. *Basidiospores* globose to shortly ellipsoid, brown, ornamented or, more rarely, smooth; trophocysts ('nurse-cells') present and often abundant, surrounding the developing spore but withering away at spore maturity. *Basidia* clavate, 2–8-spored with apical or lateral sterigmata. Terrestrial, epigeous or subhypogeous. *Type family*: *Sclerodermataceae* Corda.

Although the gleba in most cases disintegrates to produce a pulverulent spore mass, as in the true puffballs and earthstars, it is the combination of the non-development of glebal chambers lined by a hymenium, and the lack of capillitial threads that separates these orders.

Key to the British Families

1. Peridium one-layered or not distinctly stratified; fracturing irregularly to expose the gleba; fruitbody globose to tuber-like, sometimes with pseudostipe
.... **1. Sclerodermataceae**

1. Peridium complex and stratified, outer layer splitting stellately from the apex; fruitbody geastroid, sessile .... **2. Astraeaceae**

The *Glischrodermataceae* Rea, containing *Glischroderma cinctum* Fuckel, was included in the *Sclerodermatales* by Dring (1973). An account of a collection of *G. cinctum* from the Wyre Forest, Worcestershire, together with a colour plate, was published by Rea (1912), and this remains the only known British locality. Hennebert (1973), however, demonstrated the species to be a peridiate coelomycete producing conidia.

1. SCLERODERMATACEAE E. Fisch.
in Engl. & Prantl, *Nat. Pflanzenfam.* Abt. 1**, 1: 334 (1900).

Fruitbody epigeous or subhypogeous at maturity, rarely hypogeous, globose to subglobose, sessile or pseudostipitate. *Peridium* thin or thick, at times gelatinized, tough or fragile at maturity, dehiscing by irregular apical fracture. *Gleba* finally powdery, initially firm and divided irregularly by sterile tramal plates which may disrupt at maturity or harden to form 'peridioles'; capillitium absent.

Basidiospores mostly large, globose or nearly so, initially with a small pedicel, brown, inamyloid, non-cyanophilous, either smooth or with echinate to reticulate ornamentation, and a mucilaginous myxosporium. *Basidia* short clavate to pyriform, with 2–8 sterigmata. *Peridiopellis* one-layered. *Type genus*: *Scleroderma* Pers.

The family origin remains in doubt, but current views link it with the Honeycomb Truffle, *Leucogaster* Hesse, on account of spore structure. Spore development is of particular interest. Initially the spore is smooth with a short sterigmal appendage, but later it becomes surrounded by nutritional hyphae or trophocysts ('nurse-cells') within the gleba which apparently stimulate growth and at the same time are responsible for the development of the wall surface ornamentation.

Guzman (1971) recognized four genera in *Sclerodermataceae*, two of which, *Tremellogaster* E. Fischer and *Veligaster* Guzman, are restricted to the tropics.

Key to the British Genera

1. Tramal plates disintegrating at maturity, leaving a pulverulent gleba, pseudoperidioles not developed; peridium ranging from thick and corky to membranous; reportedly mycorrhizal .... **1. Earthballs** (*Scleroderma*)

1. Tramal plates hard and persistant at maturity, dividing the gleba into 'peridioles' (pseudoperidioles); peridium thin and brittle; mycorrhizal, with *Pinus* in Britain.
   .... **2. Dyeball** (*Pisolithus*)

1. EARTHBALLS

SCLERODERMA Pers.,
Synops. Meth. Fung. : 150 (1801).

Actigea Raf., *Précis Déc. Trav. Somiol.*: 52 (1814).
Sclerangium Lév. in *Ann. Sci. Nat., Bot.* sér.3, 9: 130 (1848).
Phlyctospora Corda in Sturm, *Deutschl. Fl.* 3: 41 (1851).
Pompholyx Corda in Sturm, *Deutschl Fl.* 3: 47 (1851).
Stella Massee in *Journ. Mycol.* 5: 185 (1889).

Fruitbody 1–10 cm diam., globose, reniform to tuberous, firm, fleshy, sometimes with a moderately well developed pseudostipe, with basal rhizomorphs, at times stout and numerous. *Peridium* straw-yellow to ochraceous brown, smooth, squamose to verrucose, dry, tough, dehiscing by irregular apical fracture or weathering. *Gleba* initially firm to hard, whitish, soon purple to purplish brown, interveined with very thin, white tramal plates, eventually disintegrating to form a dark purplish brown spore mass; capillitium absent. *Odour* strong, unpleasant.

Hymenium not developed; basidia formed in small clusters, spores discharged early and further development undertaken by placental hyphae and trophocysts. *Basidiospores* globose or nearly so, brown, inamyloid, non-cyanophilous, thick-walled, with fine to coarsely echinulate or reticulate ornamentation. *Basidia* clavate to pyriform, with 6–8 sterigmata. *Habitat* terrestrial or rarely on rotten wood, usually epigeous at maturity but often with subhypogeous development. *Type species*: *Scleroderma verrucosum* Pers.

Most species form mycorrhizal associations with tree species, although the fruitbodies can develop on wood. The fruitbodies are often mistaken for those of *Lycoperdon* species but lack the apical operculum for spore release, and the gleba lacks capillitial threads. The subhypogeous development, together with the dark, marbled aspect of the hard, immature gleba can also lead to confusion with the truffles (*Tuber* species). If in doubt always look for the basal attachment found on all fruitbodies. Partly owing to the latter confusion there have been many cases of commercial adulteration of foods, particularly patés, in which earthball fragments are utilized instead of the much more expensive true truffles.

It is often difficult to identify species of *Scleroderma* as they can be so variable in appearance, especially with regard to the peridial surface. Demoulin (1968) has satisfactorily defined the European species and the following key has been adapted from that provided by Demoulin & Marriott (1981).

Key to British Species

1. Spore ornament reticulate; clamp-connexions frequently present in both peridium and gleba; peridium tough, usually yellowish 2
1. Spore ornament of isolated echinulae or verrucae; clamp-connexions rare; peridium ranging from tough to membranous, yellowish to brown 4

    2. Spore ornament a complete reticulum, spores 11–14.5 μm diam. (excl. orn.); peridium thin, with smooth to minutely squamose surface; growing in sandy or well drained soil, often in gardens, parks .... **1. Potato Earthball** (*S. bovista*)
    2. Spore ornament an incomplete reticulum; peridium very thick 3

3. Peridial surface very scaly, bright yellow to orange-yellow; spores 10–13 μm diam., with crests of reticulum up to 1.5 μm high; growing on acid soil, in woodland or in boggy areas with birch; very common .... **2. Common Earthball** (*S. citrinum*)

3. Peridial surface smooth, grey or faintly yellowish; spores 8–12 μm diam., with crests of reticulum less than 1 μm high; rare, mainly Mediterranean
.... **3. Many-rooted Earthball** (*S. polyrhizum*)

    4. Peridium thick, yellow to brown, smooth or cracked, lacking a stipe-like base; spores 9–14 (–15) μm diam., plus pyramidal spines, 1–2.5 μm high; rare
.... **4. Onion Earthball** (*S. cepa*)

    4. Peridium thin, fragile at maturity, bearing numerous small squamules on a paler ground; pseudostipe more or less developed 5

5. Spores 8–11 (–13.5) μm diam., plus slender spines, 0.8–1.5 μm high; fruitbody up to 7 cm diam., with a well developed pseudostipe, up to 5 cm long; peridium reddish brown becoming yellowish brown, at first smooth soon cracking into minute irregular squamules with raised edges, individual squamules not surrounded by a ring-zone; on rich soil, often in parkland .... **5. Scaly Earthball** (*S. verrucosum*)

5. Spores (9–) 11–14 μm diam., plus slender spines 1.4–2.5 μm high; fruitbody 1.5–3 cm diam., with only a very short pseudostipe; peridium pale yellowish brown, often spotted or tinted reddish, with minute dark squamules; individual squamules leaving a ring-zone; mainly in damp localities
.... **6. Leopard-spotted Earthball** (*S. areolatum*)

1. POTATO EARTHBALL

Map No. 1

Scleroderma bovista Fr., *Syst. Mycol.* 3 : 48 (1829).
Phlyctospora fusca Corda in Sturm, *Deutschl. Fl.* 2: 51 (1841).
Scleroderma texense Berk. in *Hooker, Lond. J. Bot.* 4: 308 (1845).
Scleroderma fuscum (Corda) E. Fisch. in Engl. & Prantl, *Nat. Pflanzenfam.* 1,1: 336 (1900).
Scleroderma verrucosum Pers. subsp. *bovista* (Fr.) Sebek in *Sydowia* 7: 177 (1953).
Scleroderma verrucosum Pers. var. *bovista* (Fr.) Sebek in Pilát, *Fl. CSR, B. 1*: 570 (1958).

Selected descriptions: Coccia, Migliozzi & Lavarato (1990: 27–34); Dring (1964 : 295 as *S. fuscum*); Guzman (1970 : 338); Rea (1922 : 49); Sebek (1958 : 570).

Selected illustrations: Bon (1987 : 302); Breitenbach & Kränzlin (1986: pl.505); Dring (1964: fig. 2D–E); Ryman & Holmåsen (1984: 587); Sebek (1958: figs. 205/ 5–9).

Diagnostic characters: fruitbody with short, buried pseudostipe; peridium thin and smooth, drying fragile; spore ornament completely reticulate; clamp-connexions present.

Fruitbody 2–5 cm diam., irregularly globose, ovoid or tuberous, with a short, usually buried, longitudinally furrowed, stipe-like base, 2–4 cm long, and a fibrillose mycelium. *Peridium* thin, 1–1.5 mm thick, drying fragile and brittle, pale yellowish to greyish brown, with yellow tones persisting on the underside, smooth or often developing minute, innate squamules. *Gleba* at first white, becoming purplish black, marbled with fine, yellowish tramal plates, greyish brown to blackish brown and pulverulent at maturity.
   Clamp-connexions present. *Basidiospores* 11–14.5 μm diam. (excluding ornament), dark brown, with the very dark brown ornament forming a complete reticulum, 1–2 μm high; usually retaining remnants of trophocysts attached to the ornament.

Habitat: on well-drained, often sandy soil, especially along roadsides in parks and gardens, often associated with lime (*Tilia*) but also with many other trees.

Distribution and frequency: fairly common throughout the British Isles; widespread throughout the North Temperate zone.

Other remarks: Often regarded in the past as a variety of the Scaly Earthball (*S. verrucosum*) by virtue of the thin peridium, but the presence of clamp-connexions and the reticulated spores indicate there is no such relationship. The young spores often retain the trophocysts which form a 'multicellular envelope', a character on which *Phlyctospora fusca* was based.

Potato Earthball

Fig. 6. *Scleroderma bovista* (Denmark, Århus, Lake Braebrand, 29 Aug.1989, C. Lange).

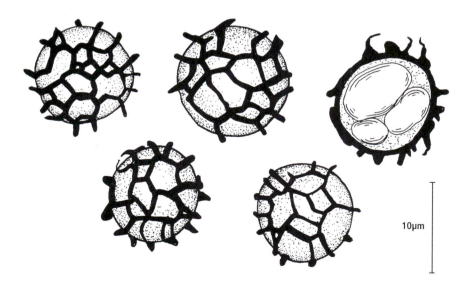

Fig. 7. *Scleroderma bovista* (Lancashire, Freshfield, 14 Sept. 1963, Kotlaba & Palmer 11770). Spores.

2. COMMON EARTHBALL

Map No. 2

Scleroderma citrinum Pers.: Pers., *Synops. Meth. Fung.* : 153 (1801).
Lycoperdon tessulatum Schumach., *Enum. Plant. Part Saell.* 2: 191 (1802).
Scleroderma vulgare Hornem. in *Fl. Danica*: pl.1969/2 (1829).
Scleroderma aurantium sensu Rea (1922), Ramsbottom (1953), Wakefield & Dennis (1981) non *Lycoperdon aurantium* L., *Spec. Plant.*: 1053 (1753).

Selected descriptions: Guzman (1970: 329); Massee (1889: 49, fig. 45).

Selected illustrations: Breitenbach & Kränzlin (1986: pl.506); Coccia, Migliozzi & Lavarato (1990 : 47–51); Guzman (1970 : pl.1 and 5); Phillips (1981: 250); Ramsbottom (1953: pl.32b); Ryman & Holmåsen (1984: 586); Sarasini (1991: 123); Wakefield & Dennis (1981: 207, pl.111/3).

Diagnostic characters: thick peridium (–5 mm) coarsely squamose, often bright yellow; never forming a pseudostipe; spores 10–13 μm diam., with incomplete reticulum; clamp-connexions present.

Fruitbody 5–10 (–15) cm diam., subglobose, ovoid, sometimes tuberous, often apically flattened but lacking a stipe-like base, arising from white mycelial cords; gregarious and caespitose. *Peridium* 2–5 mm thick, tough, whitish in section, pale brown tinged yellow to bright orange-yellow, with large, coarse scales, drying hard. *Gleba* greyish white then purplish black to black, marbled with thin, white tramal plates, remaining firm for a long time, finally greyish olive-brown and pulverulent.

Clamp-connexions present. *Basidiospores* 10–13 μm diam. (excluding ornament), globose, dark fuscous brown, with a surface ornament forming an incomplete reticulum of crests, 0.8–1.5 μm high. *Basidia* 5–10 x 3–5 μm, clavato-pyriform, bearing 2–5 sterigmata of unequal length.

Habitat: on acid soils in woodland and heathland, often associated with beech (*Fagus*), oak (*Quercus*) and birch (*Betula*), especially on sandy soils, from late summer to early autumn.

Distribution and frequency: widespread and common throughout the British Isles. Very common and widely distributed throughout North Temperate zone, especially western Europe and eastern North America.

Other remarks: The most common species of *Scleroderma*, frequently cited under the names *S. aurantium* and *S. vulgare*. Sometimes associated with the Parasitic Boletus, *Xerocomus parasiticus* (Bull.: Fr.) Quél.

Fig. 8. *Scleroderma citrinum* (Denmark, Jutland, 1993, Spooner).

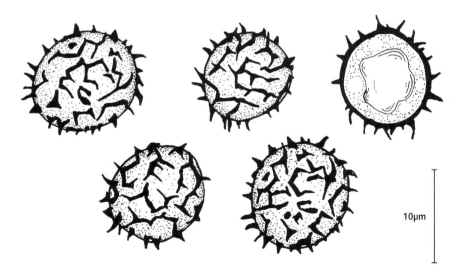

Fig. 9. *Scleroderma citrinum* (Surrey, Brooklands, 22 Sept. 1957, Balfour-Browne). Spores.

3. MANY-ROOTED EARTHBALL

Map No. 3

Scleroderma polyrhizum (J. F. Gmel.) Pers., *Synops. Meth. Fung.*: 156 (1801).
Lycoperdon polyrhizum J. F. Gmel., *Syst. Nat. Linn.* 2: 1464 (1796).
Scleroderma geaster Fr., *Syst. Mycol.* 3 : 46 (1829).
Sclerangium polyrhizum (Pers.) Lév. in *Ann. Sci. Nat., Bot.* 3, 9: 130 (1848).
Stella americana Massee in *J. Mycol.* 5: 185 (1890).

Selected descriptions: Broome (1870); Coccia, Migliozzi & Lavarato (1990: 42–47); Guzman (1970: 312); Rea (1922 : 50).

Selected illustrations: Léveillé (1848 : pl.7); Massee (1889 : fig.35); Sarasini (1991: 124); Sebek (1958: fig. 203); W. G. Smith in Broome (1870).

Diagnostic characters: large fruitbodies; thick peridium (–5 mm) smooth and greyish, apically splitting in stellate manner; spores 8–10 µm diam., with low ornament of an incomplete reticulum; clamp-connexions present.

Fruitbody 4–15 (–18) cm diam., globose to tuberous or irregularly lobed, tapering with a compact mycelial base but lacking a pseudostipe. *Peridium* very thick, up to 5 mm, pale yellowish, greyish brown when old, smooth then disrupting into small, irregular squamules; finally splitting at the apex into recurved unequal lobes. *Gleba* purplish umber, finally rusty brown to sepia-brown.
    Clamp-connexions present. *Basidiospores* 8–12 (–14) µm diam. (excluding ornament), globose, dark fuscous brown, with a low surface ornament of small verrucae forming an incomplete reticulum, with crests 0.4–0.8 µm high.

Habitat: prefers sandy soil, in deciduous woodland.

Distribution and frequency: rare; British records largely confined to south-east England and the Welsh borders, and parts of Ireland. Known from southern North America and from Europe, particularly around the Mediterranean region.

Other remarks: more frequently cited in the literature as *S. geaster*. A large and robust, Mediterranean-climate species, occasionally found in England. The stellate dehiscence of the peridium resulted in Léveillé (1848) proposing the genus *Sclerangium* for this species. The spore structure would suggest a relationship closest to *S. citrinum*. Could be confused with an earthstar (*Geastrum* sp.).

Fig. 10. *Scleroderma polyrhizum* (France, Chiavari, 14 Oct. 1972, Demoulin 4553).

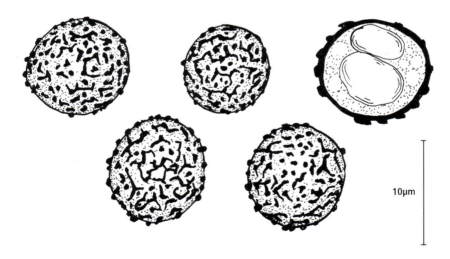

Fig. 11. *Scleroderma polyrhizum.* (Hereford, Oct. 1870, Broome). Spores.

4. ONION EARTHBALL

Map No. 4

Scleroderma cepa Pers.: Pers., *Synops. Meth. Fung.*: 155 (1801).
Scleroderma cepioides Gray, *Nat Arrang. Brit. Pl.* 1: 582 (1821).
Scleroderma vulgare Hornem. var. *spadiceum* sensu W. G. Smith, *Synops. Brit. Basid.*:
479 (1908).
Scleroderma spadiceum sensu Rea (1922), non *S. spadiceum* Schaeffer: Pers., *op. cit.*: .
155 (1801, nom. confusum).

Selected descriptions: Coccia, Migliozzi & Lavarato (1990 : 9–14); Demoulin (1967:
301); Rea (1922 : 50); Sebek (1958 : 805 as *S. aurantiacum* var. *spadiceum*); Wakefield
& Dennis (1950: 274).

Selected illustrations: Sebek (1958 : fig. 201/4-6); Vaillant (1727: 123, pl.16/5-6);
Wakefield & Dennis (1950: pl.111/4B).

Diagnostic characters: fruitbody lacking pseudostipe; peridium very thick, smooth,
finally cracking; spores with isolated verrucae; clamp-connexions absent.

Fruitbody 2–6 cm diam., subglobose, pyriform or flattened, lacking a stipe-like base but
sometimes tapering below. *Peridium* thick and rigid when fresh, drying thinner, cream-
yellow to dark reddish brown, at times with yellow or orange tints on drying, smooth but
eventually coarsely cracked with minute squamules above, with irregular apical
dehiscence. *Gleba* at first whitish soon fuscous brown with violaceous tints, finally
greyish brown to olivaceous and pulverulent.
  Clamp-connexions absent. *Basidiospores* 9–14 (–15) μm diam. (excluding ornament),
globose, dark brown, thick-walled, with a surface ornament of isolated, pyramidal
verrucae, 1–2.5 μm high and about 1.0 μm broad at their base; often surrounded by
numerous, thin-walled, hyaline trophocysts.

Habitat: prefers sandy ground, often under species of oak (*Quercus*).

Distribution and frequency: uncommon but widespread in Britain; occurs throughout
western Europe and North America.

Other remarks: the smooth peridial surface is vaguely reminiscent of a red-skinned
onion, hence the epithet, 'cepa'. Often regarded as a variety of *S. verrucosum*. Although
the spores are verrucose, the surface ornament often retains numerous trophocysts.

Onion Earthball

Fig. 12. *Scleroderma cepa* (Sussex, Chichester, 18 Sept. 1987, Goodchild).

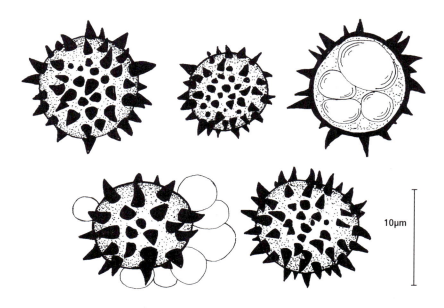

Fig. 13. *Scleroderma cepa.* (Norfolk, Honingham, 8 Oct. 1958, Ellis). Spores.

5. SCALY EARTHBALL

Map No. 5

Scleroderma verrucosum (Bull.: Pers.) Pers., *Synops. Meth. Fung.*: 154 (1801).
Lycoperdon verrucosum Bull., *Herb. Fr.*: 157 (1791).

Selected descriptions: Guzman (1970 : 276); Massee (1889 : 50); Rea (1922 : 50); Sebek (1958 : 567).

Selected illustrations: Bulliard (1781: pl.24); Coccia, Migliozzi & Lavarato (1990: 20–24); Gerhardt (1985: 188); Greville (1823: pl.48); Phillips (1981: 250); Ryman & Holmåsen (1984: 587); Sarasin (1991: 123); Sebek (1958: figs. 204/ 1–3, 205); Wakefield & Dennis (1981: 208, pl. 111/4).

Diagnostic characters: fruitbody with well developed, furrowed pseudostipe; peridium thin, finally fragile, with minute scales lacking ring-zones; spores 9–11 µm diam., with isolated echinulae; clamp-connexions absent.

Fruitbody 2–5 (–8) cm diam., globose or apically depressed, occasionally almost sessile but usually arising from a stipe-like base, up to 5 (–10) cm long, which is longitudinally ridged, with basal, white mycelial cords. *Peridium* relatively thin, 0.5–1 mm thick, leathery but fragile when dry, reddish brown becoming ochraceous brown, darkening with age, surface bearing numerous, small, angular, innate squamules, becoming smooth; individual squamules not surrounded by a ring-zone. *Gleba* at first cream-coloured becoming purplish black, marbled with fine, white tramal plates, finally pulverulent and greyish brown.
    Clamp-connexions absent. *Basidiospores* 9–11 (–13.5) µm diam. (excluding ornament), globose, fuscous brown, with surface bearing slender, isolated echinulae, 0.8–1.5 µm high.

Habitat: On both rich and sandy soil, in deciduous woodlands, parkland and heathlands, occasionally on rotting wood.

Distribution and frequency: common and widespread in the British Isles and North and South Temperate zones.

Other remarks: in spite of the name, the peridium is usually less scaly than that found in the Common Earthball (*S. citrinum*).

Scaly Earthball

Fig. 14. *Scleroderma verrucosum* (Devon, Ashclyst Forest, 20 Sept. 1986, Roberts).

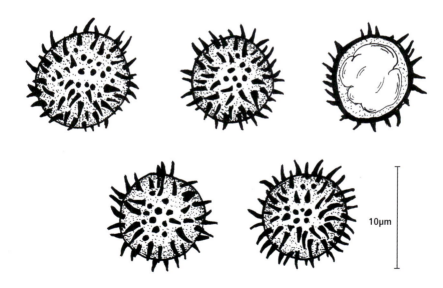

Fig. 15. *Scleroderma verrucosum*. (Surrey, Dorking, 4 Oct. 1990, Legon). Spores.

6. LEOPARD-SPOTTED EARTHBALL

Map No. 6

Scleroderma areolatum Ehrenb., *Sylvae Mycol. Berol.* 15: 27 (1818).
Scleroderma lycoperdoides Schwein., *Schrift. Natürf. Ges. Leipzig* 1 : 61 (1822).
Scleroderma verrucosum Pers. var. *violascens* Herink ex Sebek in *Sydowia* 7 : 176 (1953).

Selected descriptions: Coccia, Migliozzi & Lavarato (1990: 24–27); Guzman (1970: 282)

Selected illustrations: Breitenbach & Kränzlin (1986: pl. 504); Gerhardt (1985: 189); Phillips (1981: 250).

Diagnostic characters: fruitbody small, with very short pseudostipe; peridium thin, bearing minute scales with ring-zones; spores large, 11–14 μm diam., with isolated echinulae.

Fruitbody small, 1.5–3 (–4) cm diam., subglobose, pyriform or tuberous, tapering below with a very reduced, buried pseudostipe, which is longitudinally furrowed, and bears a few mycelial strands. *Peridium* thin, fragile at maturity, pale yellowish brown, spotted reddish brown, or sometimes developing pink, violaceous or reddish tints, bearing numerous, small, dark brown squamules on a paler ground; individual squamules surrounded by a ring-zone leaving a reticulate pattern after squamules are worn away. *Context* whitish or often discoloured flesh-pink to purplish. *Gleba* at first white to cream-coloured then deep purplish brown, marbled, finally greyish brown and pulverulent.
   Clamp-connexions absent. *Basidiospores* (9–) 11–14 (–15) μm diam.(excluding ornament), globose, dark brown, with surface ornament of isolated echinulae, 1.4–2.5 μm high.

Habitat: solitary or gregarious, amongst grass at edge of woods and fields, possibly mycorrhizal with oak (*Quercus*), preferring damp localities.

Distribution and frequency: widespread but occasional throughout the British Isles; frequent throughout northern Europe, less common towards the Mediterranean; common in North America, and possibly in south temperate zone.

Other remarks: very similar macroscopically to *S. verrucosum*, but with significantly larger spores.

Fig. 16. *Scleroderma areolatum* (Laessøe).

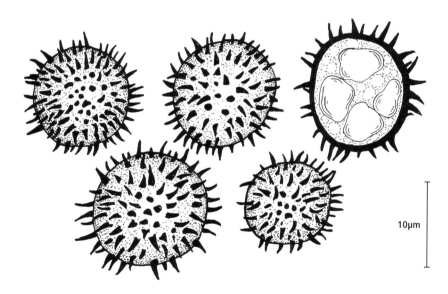

Fig. 17. *Scleroderma areolatum*. (Essex, Hadleigh, Belfair NR, 29 Aug. 1987, Spooner). Spores.

PISOLITHUS Alb. & Schwein.,

Consp. fung. Lusatiae sup. : 82 (1805).

Only species: *Pisolithus arhizus* (Scop. : Pers.) Rausch.

DYEBALL

Map. No. 7

Pisolithus arhizus (Scop. : Pers.) Rausch. in *Zeitschr. Pilzk.* 25: 50 (1959).
Scleroderma arhizum Scop.: Pers., *Synops. Meth. Fung.*: 152 (1801).
Polysaccum pisocarpium Fr., *Syst. Mycol.* 3: 54 (1829).
Polysaccum tinctorium (Mich. : Pers.) Mont., *Phyt. Canar.*: 87 (1840).
Pisolithus tinctorius (Mich. : Pers.) Coker & Couch, *Gasterom. Eastern US & Canada*: 170 (1928).

Selected descriptions: Marchand (1973: 174); Pilat (1958: 576); Ramsbottom (1953: 175); Rea (1922: 50).

Selected illustrations: Gerhardt (1985: 190); Hansen (1986: Figs. 3–4); Marchand (1973: pl.183); Massee (1889: fig. 53); Pegler (1990: 169); Phillips (1981: 251); Sowerby (1814: pl.425 a–b).

Diagnostic characters: large fruitbodies; yellow mycelium; multilocular gleba.

Fruitbody with a subglobose head, 5–9 (–11) cm diam., prolonged into a solid and firm , irregular pseudostipe, 2–3 cm diam. sometimes more, and varying from 1–8 cm long, incorporating soil particles, with greenish yellow mycelial cords. *Peridium* very thin, about 1 mm thick, brittle and dry; at first whitish then yellowish ochraceous with olive-black patches, smooth, eventually cracking and flaking away at the apex to expose the pseudoperidioles. *Gleba* divided, with the tramal plates forming individual pseudoperidioles, about 5 x 2 mm, which are subspherical and usually compressed, at first whitish then reddish brown to almost black, becoming pulverulent; pseudoperidioles larger and more mature to the peridial apex.
    Clamp-connexions present. *Basidiospores* 7–10 μm diam. (excluding ornament), globose, thick-walled, with long, curved spines, 1.0–2.3 μm high, dark yellowish brown.

Habitat: in warm, dry sites, on disturbed land and roadsides; amongst sand or gravel, but never on calcareous soils. Forming mycorrhizal associations with pine in Britain. Occurring from spring to autumn.

Distribution and frequency: very rare in Britain, with records from London, Hampshire, Devon, Norfolk and Republic of Ireland (Tipperary). Almost cosmopolitan except in cold temperate and polar regions.

Other remarks: *Pisolithus* was almost certainly introduced to Europe, probably with *Eucalyptus* from Australasia. An account of *Pisolithus* in Ireland was published by Ing (1985). Very variable in form, from subsessile to long-stemmed, resulting in many epithets. The hard sterile, subterranean pseudostipe can remain for several months. The epithet '*tinctorius*' refers to the use of the fungus for dyeing wools; a bright olivaceous yellow pigment is contained in the gelatinous layers of the pseudoperidiole wall.

Dyeball

Fig. 18. *Pisolithus arhizus* (Siberia, Trans-Baikal, Christensen).

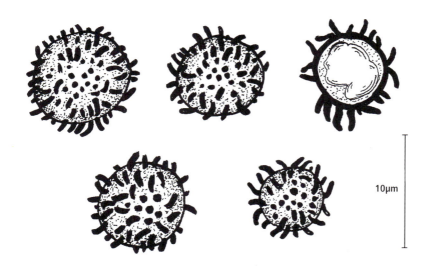

Fig. 19. *Pisolithus arhizus*. (Surrey, Windsor Great Park, 3 Oct. 1956, Green). Spores.

British Puffballs, Earthstars and Stinkhorns

2. ASTRAEACEAE Jülich
Bibl. Mycol. 85: 305 (1981).

Astraeaceae G.W. Martin in *Stud. Nat. Hist. Iowa Univ.* 17: 103 (1936) (nom. inval.)
Astraeaceae Zeller in *Mycologia* 41: 53 (1949) (nom. inval.)
Astraeaceae V.J. Stanek in Pilát, *Flora CSR* B. 1: 626 (1958) (nom. inval.)

Fruitbody epigeous at maturity, geastroid, subglobose or depressed, sessile, comprising a layered exoperidium which splits stellately into rays at maturity to expose an endoperidial body. *Rays* hygroscopic. *Endoperidial body* subglobose, opening by an apical tear, peristome absent. *Endoperidium* thin, with felty surface. *Gleba* white at first, divided into chambers by thin tramal plates, dark brown at maturity, powdery; paracapillitium present, comprising hyaline, clamped, usually thick-walled, branched hyphae.
  Hymenium not developed. *Basidiospores* dark brown, globose, inamyloid, echinulate or verruculose. *Basidia* broadly clavate to capitate, with 2–8 sterigmata. *Type genus*: *Astraeus* Morgan.

The family was proposed, though not validly published, for the single genus *Astraeus* by Martin (1936), and was referred to the *Sclerodermatales* by Dring (1973). This position has been generally accepted and was endorsed by Bronchart et al. (1975), who examined the ultrastructure of the basidiospore wall. These authors showed that the inner spore wall of *A. hygrometricus* consists of alternating electron-dense and electron-transparent layers, unlike that of species of *Geastrum*, but similar to that of species of *Scleroderma* and *Pisolithus* (*Sclerodermatales*). *Astraeus* was referred by Morgan (1889) to *Lycoperdaceae*, and later placed in *Calostomataceae* by Fischer (1900). That this position was unsatisfactory was suggested by Coker & Couch (1928), who also considered that a new family was required to accommodate the genus. *Myriostoma* and *Endogonopsis* Heim were also referred to *Astraeaceae* by Dring (1973). *Myriostoma*, as noted below, proves to be better placed in *Geastraceae*; *Endogonopsis*, based on a single Indian species, requires elucidation. It may represent just young or abnormal development of *Astraeus*.

ASTRAEUS Morgan,
J. Cincinnati Soc. Nat. Hist. 12: 19–20 (1889).

Fruitbody 1.5–6 cm diam., sessile, with basal tuft of hyphae and arising from delicate rhizomorphs. *Odour* not distinctive. *Exoperidium* thick, stratified, hard and woody when dry. *Endoperidial body* subglobose, opening by an irregular, apical tear. *Columella* lacking. *Paracapillitium* hyaline, branched, mostly thick-walled, with clamp-connexions. *Basidia* capitate to broadly clavate, bearing 2–8 sessile spores. *Basidiospores* globose, 7–11 µm diam. (excluding ornament), dark brown, echinulate or verruculose, wall with 2–4 alternating electron-dense and electron-transparent layers (Sunhede, 1989). **Habitat** on soil, occurring in mycorrhizal association with woody plants. **Distribution** cosmopolitan. *Type species*: *Astraeus hygrometricus* (Pers.) Morgan.

Fruitbodies of this genus closely resemble those of species of *Geastrum*, having an endoperidial body which is exposed at maturity by splitting of the exoperidium into rays which turn downwards to form a star-like pattern. However, the genus differs from

Geastrum morphologically in several important respects, notably in lacking a columella, in lacking a defined peristome, and in having branched capillitial hyphae which bear clamp connexions. It also differs in the form of the basidia, in the comparatively large spores, and in the structure of the spore wall. Two other species have been recognized in the genus: *A. pteridis* (Shear) Zeller, from western U.S.A., which differs from the type in its much larger fruitbodies, and the Asian *A. koreanus* (Stanek) Kreisel, differing in its comparatively small, pale and fragile fruitbodies.

There is a single British species.

BAROMETER EARTHSTAR

Map No. 8

Astraeus hygrometricus (Pers.) Morgan in *J. Cincinnati Soc. Nat. Hist.* 12: 20 (1889)
Geastrum hygrometricum Pers., *Synops. Meth. Fung.*: 135 (1801).
Geastrum fibrillosum Schwein., *Schriften Naturf. Ges. Leipzig* 1: 59 (1822) (*fide* Coker & Couch 1928).
Lycoperdon stellatum Scop., *Fl. carniol.* ed. 2, 2: 489 (1772); [ed 1:63 (1760)].
Astraeus stellatus (Scop.) E.Fisch., *Nat. Pflanzenfam.* 1, 1: 341 (1900).

Selected descriptions: Bon (1987: 302); Coker & Couch (1928: 185 -188)

Selected illustrations: Bon (1987: 303); Cetto (1988: 368); Chaumeton (1987: 380); Coker & Couch (1928: pl. 31, 77, 115 fig. 37, 117 figs 4–7); Cooke (1871: 371); Dörfelt (1985: figs 25, 26); Ellis (1981: figs 1/1–3 & 3/1); Gerhardt (1985: 191); Marchand (1976: pl. 355); Miller & Miller (1988: pl. VII B, fig. 36); Moser & Jülich (1988: VII, 1 & 2); Phillips (1981: 254); Rea (1912: pl. 17).

Diagnostic characters: fruitbody geastroid, exoperidium thick, splitting into rays, strongly hygroscopic; endoperidial body sessile; peristome lacking; spores 7.5–11 µm diam. (excluding ornament), minutely echinulate.

Fruitbody often gregarious, partly hypogeous in development, epigeous at maturity, subglobose or depressed, 2–3.5 cm diam., with basal mycelium, surface often partly encrusted with debris. Expanded fruitbody 5–9 cm across, exoperidium splitting to usually more than halfway into 7–13 rays, strongly hygroscopic, rays arched downwards when damp, turned upwards to cover the endoperidial body when dry. *Exoperidium* fleshy, distinctly layered; outermost layer felty, pale buff, lost with age; fibrous layer whitish or yellowish becoming dark brown, thicker towards the base, overlying a thinner, horny layer. *Pseudoparenchymatous layer* yellowish brown to dark brown, up to c. 1.5 mm thick, soon splitting into a characteristic reticulate pattern, hard when dry. *Endoperidial body* sessile, subglobose, 1.5–3 cm diam. *Endoperidium* felty, whitish at first, becoming pale buff. *Peristome* lacking, endoperidial body opening by an irregular, often longitudinal, apical tear. *Columella* lacking. *Mature gleba* brown.
   Basidiospores globose, brown, minutely echinulate or verruculose, 7.5–10 (–11) µm diam. excluding ornament, 9–11.5 (–12.5) µm diam. including ornament, verruculae tapered, irregular, 0.3–0.7 µm high. *Basidia* short clavate, with very short sterigmata. *Paracapillitium* hyaline, 3–6 (–9) µm diam., branched, with clamp-connexions, walls thin or thickened with a narrow, sometimes discontinuous lumen, non-sepate, surface not or slightly encrusted.

Habitat: on nutrient-poor, usually sandy soil, preferring dry conditions; often in wooded areas, occurring in mycorrhizal association with various trees.

Distribution and frequency: uncommon in Britain and occurring mainly in southern England. Cosmopolitan excluding arctic, alpine and cold temperate (boreal) regions.

Other remarks: the species should be easily recognisable by the thick, hygroscopic exoperidium which usually splits into numerous, narrow rays. It is regarded as an adaptation to dry habitats.

Barometer Earthstar

Fig. 20. *Astraeus hygrometricus* (Laessøe).

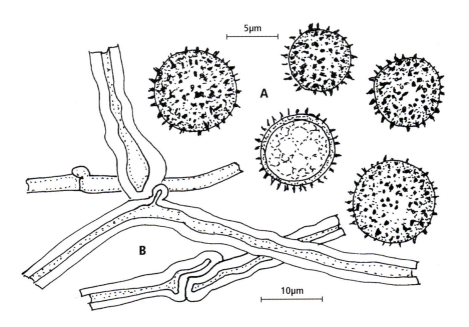

Fig. 21. *Astraeus hygrometricus*. (Kent, Oct. 1991, Weightman). A, Spores; B, paracapillitial threads.

STALK AND STILT PUFFBALLS

TULOSTOMATALES Demoulin
in *Bull. Jard. Bot. Belg.* 38: 9 (1968).

Fruitbody initially subglobose, usually with hypogeous development, epigeous at maturity, with a globose or subglobose head borne on a well-developed stipe, sometimes with a volva. *Stipe* composed of thick-walled, longitudinally arranged hyphae, interwoven at centre, dry or rarely gelatinous. *Peridium* with two layers, dehiscent either by an apical pore or pores, or by irregular lobes, or with circumscissile cleavage. *Gleba* not loculate, powdery at maturity.

Capillitium or *Paracapillitium* present. *Basidia* neither palisadic nor hymenial. *Basidiospores* globose to broadly ellipsoid, yellowish brown, smooth or ornamented.

Three families can be recognized. *Calostomataceae* Pilát, with a single genus *Calostoma* Desv., is characterized by the complex, 4-layered peridium. Species of *Calostoma* are mostly tropical or occur in warm-temperate regions and are not known from northern Europe.

Key to British Families

1. Peridial dehiscence either by a pore or by irregular lobes; capillitial threads smooth .... **1. Stalk Puffballs** (*Tulostomataceae*)
1. Peridial dehiscence by circumscissile splitting; capillitial threads with spiral or annulate ornament .... **2. Stilt Puffballs** (*Battaraeaceae*)

1. TULOSTOMATACEAE E. Fisch.
in *Nat. Pflanzenfam.* 1**, 1: 342 (1900).

Fruitbody with a 2-layered peridium. *Stipe* cylindrical, usually well-developed, smooth or scaly, sometimes with a basal bulb, apically inserted into a socket at the base of the endoperidial body. *Exoperidium* membranous or hyphal, sometimes fugacious. *Endoperidium* membranous, dehiscing by a well-defined apical pore or by breaking down into patches or splitting into irregular lobes which turn outwards. *Gleba* brown.

Capillitium well-developed, with simple or branched, hyaline to yellowish or brown hyphae, often thick-walled, septate, sometimes swollen at the septa. *Basidia* with lateral sterigmata, borne in nests evenly scattered throughout the gleba (Jülich 1981). *Basidiospores* with verruculose or reticulate ornament. *Type genus*: *Tulostoma* Pers.: Pers.

The family is worldwide in distribution. Six genera are recognized: *Dictyocephalus* Underw., *Chlamydopus* Speg., *Schizostoma* Ehrenb. are unknown in Europe, and *Phellorina* Berk. is known from the Mediterranean region. Many species of *Tulostomataceae* are adapted to arid conditions, though *Tulostoma* includes species from a wide range of habitats, including tropical rain forest (Wright 1987).

Key to British Genera of Tulostomataceae

1. Fruitbodies dehiscing by an apical peristome; slender, endoperidial body (in British species) rarely over 1 cm diam .... **1. Stalk Puffballs** (*Tulostoma*)
1. Fruitbodies dehiscing by irregular breakdown of endoperidium; robust, endoperidial body 2.5–4 cm diam .... **2. Quelet's Stalk Puffball** (*Queletia*)

1. STALK PUFFBALLS

TULOSTOMA Pers.: Pers.,

Synops. Meth. Fung.: 139 (1801); *Römers Neues Mag. bot.* 1: 86 (1794).

Tylostoma Spreng., *Syst. Veget.* IV (1): 378 (1797).
Tulasnodea Fr., *Summa Veg. Scand.*: 440 (1849).

Fruitbody usually hypogeous in development, epigeous and stipitate at maturity. *Stipe* cylindric, whitish to yellowish brown, fibrous or smooth, inserted into a socket at the base of the endoperidial body. *Exoperidium* membranous or hyphal and inconspicuous, not persistent. *Endoperidial body* subglobose or depressed, sometimes ovate, usually with a single, apical peristome, commonly concave beneath forming around the stipe apex a socket which may be fringed with lacerate remains of the exoperidium. *Endoperidium* white to yellowish or brown, velvety, pruinose or smooth. *Peristome* variable, usually circular, well-defined and raised in British species, sometimes darker than the surrounding endoperidium. *Gleba* yellowish brown, powdery at maturity.

Capillitial hyphae branched, hyaline or pigmented, septate, without clamp-connexions, septa simple or inflated, commonly pigmented, walls mostly thickened, lumen often narrow, sometimes discontinuous, surface smooth or slightly encrusted. *Basidia* simple, narrowly clavate, terminal or lateral, 4-sterigmate (Coker & Couch 1928). *Basidiospores* globose or subglobose, ornamented or rarely smooth, yellowish-brown, mostly 4–7 μm diam. *Type species*: *T. brumale* Pers.: Pers.

This is a large genus of temperate and tropical species. Species are delimited mainly on characters of the peristome, exoperidium, capillitium and spores. A few species, including *T. brumale*, are reported as mycorrhizal (see Wright 1987).

Only three species are known from the British Isles, as indicated in the key. *Tulostoma fimbriatum* Fr. has also been reported as British by Pouzar (1958), followed by Jülich (1984), but there seems to be no evidence for this. There is no such material at Kew and, according to Palmer (1968), the basis of this record cannot now be traced.

Key to British Species

1. In moss tussocks on limestone rocks; basal mycelium white, conspicuous; endoperidium white; stipe smooth; capillitial hyphae 2–6 μm diam.
   .... **1. White Stalk Puffball** (*T. niveum*)

1. In sandy soil; basal mycelium binding sand, not white and conspicuous; endoperidium coloured, cream to yellowish brown; stipe surface fibrous; capillitial hyphae commonly up to 9 μm diam. 2

2. Spores < 5 μm diam. excluding ornament; septa of capillitial hyphae mostly conspicuously inflated, to 14 μm diam., and distinctly pigmented. Exoperidium membranous; socket with lacerate membrane
   .... **2. Winter Stalk Puffball** (*T. brumale*)

2. Spores 5–6.5 μm diam. excluding ornament; septa of capillitial hyphae not or slightly inflated, hyaline or yellowish. Exoperidium hyphal, inconspicuous; socket not prominent, scaly .... **3. Scaly Stalk Puffball** (*T. melanocyclum*)

1. WHITE STALK PUFFBALL

Map No. 9

Tulostoma niveum Kers in *Bot. Not.* 131: 411 (1978).

Selected descriptions: Kers (1978: 411–417); Wright (1987: 162–164).

Selected illustrations: Kers (1978: figs 1,2,5); Wright (1987: fig. 105, pl. 6/5–6, pl. 46/3).

Diagnostic characters: small, white, fragile fruitbodies; coarsely verruculose spores 5–6.5 μm diam. (excluding ornament); habitat in moss tussocks on limestone rocks.

Fruitbody stipitate at maturity, up to c. 25 mm high, stipe deeply immersed in moss, arising from copious white mycelium. *Stipe* slender, sometimes short, usually elongating at maturity, white or yellowish, 0.5–2 mm diam., smooth or striate. *Exoperidium* membranous, white, not binding debris. *Endoperidial body* 4.5 (–9.5) mm diam., subglobose or ovate, papillate, with a single peristome, base flattened, with slight collar around the stipe apex. *Endoperidium* white, becoming yellowish particularly towards the base, slightly velvety, becoming smooth. *Peristome* obtusely conical, white, aperture circular. *Gleba* pale brown or fawn.

Capillitial hyphae 2–6 μm diam., hyaline, branched, usually thick-walled, lumen often narrow, sometimes discontinuous, septate, septa not or slightly enlarged, often yellowish, slightly constricted, 5–7 μm diam., surface smooth or irregularly slightly encrusted. *Basidia* not seen. *Basidiospores* globose or subglobose, pale yellowish, verruculose, 5–6 (–6.5) μm diam. excluding ornament, 5.6–7.5 μm diam. including ornament, verruculae somewhat irregular in size and form, often coarse, mostly isolated, 0.2–0.6 μm high, 0.2–1.6 μm across.

Habitat: saprophytic and perennial in thick moss tussocks on bare limestone rocks in treeless areas.

Distribution and frequency: in Britain known from only a single locality, near Inchnadamph in Scotland. Known otherwise only from Sweden.

Other remarks: The species is distinctive, both for its small, fragile, pure white fruitbodies and for its habitat. It fruits from early winter to early spring and the fruitbodies soon collapse and disintegrate thereafter (Kers, 1978), being rarely present later in the year. The above description is based on a single collection: Scotland, 1.5 km S. of Inchnadamph, in moss on a limestone boulder, 10 May 1992, C. Scouller (K).

White Stalk Puffball

Fig. 22. *Tulostoma niveum*. (Sweden, Borgholm, 6 Oct. 1986, Nitare).

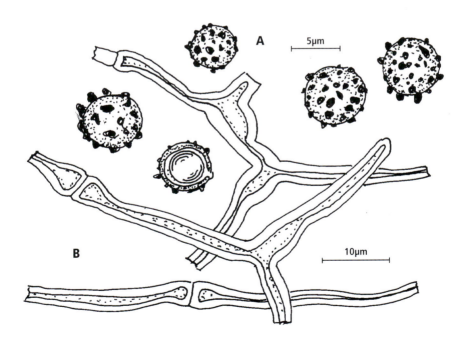

Fig. 23. *Tulostoma niveum*. (Scotland, May 1992, Scoutter) A, Spores; B, paracapillitial threads.

2. WINTER STALK PUFFBALL

Map No. 10

Tulostoma brumale Pers.: Pers., *Synops. Meth. Fung.*: 6 (1801); *Römers Neues Mag. bot.* 1: 86 (1794).
Lycoperdon mammosum P. Micheli, *Nova Plant. Gen.*: 217 (1729) *fide* Wright, 1987).
Tulostoma mammosum (P. Micheli) Fr., *Syst. Mycol.* 3: 42 (1821).
Tulasnodea mammosa (P. Micheli) Fr., *Summa Veg. Scand., Sect. Post.*: 440 (1849).
Lycoperdon pedunculatum L., *Sp. Pl.* 2: 1654 (1763).
Tulostoma pedunculatum (L.) J.Schröt. in *Beitr. Biol. Pflanzen* 2: 65 (1877).
Tulostoma brevipes Petri in *Ann. Mycol.* 2: 418 (1904) *fide* Wright 1987.
Tulostoma fuscoviolaceostipitatum Shvarm. & N.M.Filim., *Fl. Spor. Rast. Kazhakstan, VI: Gasteromycetes*: 227 (1970) *fide* Wright, 1987.

Selected descriptions: Breitenbach & Kränzlin (1986: 396); Phillips (1981: 251); Wright (1987: 76–80).

Selected illustrations: Breitenbach & Kränzlin (1986: 397); Cetto (1988: 379); Michael, Hennig & Kreisel (1986: fig 175); Mornand (1989: fig. 2); Phillips (1981: 251); Pouzar (in Pilát 1958: fig. 213); Wright (1987: figs 23–24; pl. III figs 1–4, XXXII fig. 2).

Diagnostic characters: spores sparsely verruculose, 4.5–5 µm diam. (excluding ornament); septa of capillitial hyphae inflated, pigmented; exoperidium membranous, socket with lacerate membrane.

Fruitbody epigeous at maturity, stipitate, 1.8–3.5 cm high, binding sand at base; solitary or gregarious. *Stipe* cylindric, 1.5–2.5 mm diam., pale yellowish or straw, darker yellow-brown below, surface smooth to fibrous, sometimes splitting to form recurved scales. *Exoperidium* membranous, soon lost. *Endoperidial body* 6–10 mm diam., with a single peristome, subglobose, depressed beneath or not, with an irregular, lacerate, collar-like membrane usually present around the stipe apex. *Endoperidium* pale yellowish or straw, smooth. *Peristome* well defined, circular, usually somewhat raised and often darker than the surrounding endoperidium. *Gleba* pale yellowish.
   Capillitial hyphae (2.5–) 3–9 µm diam., branched, septate, hyaline or pale yellowish-brown, more deeply pigmented at septa, thick-walled, with irregular, usually narrow, often discontinuous lumen, expanded at septa to 5–14 µm diam., surface smooth or partly encrusted. *Basidia* not seen. *Basidiospores* 4.2–4.8 (–5.2) µm diam. excluding ornament, 4.4–5.1 (–5.5) µm diam. including ornament, verruculae isolated, c. 0.1–0.2 µm high, globose or subglobose, pale yellowish brown.

Habitat: usually on sandy, calcareous soil, amongst grass and herbs, typically in dune slacks.

Distribution and frequency: uncommon in the British Isles, but the most common species of *Tulostoma* in western Europe. In Britain, it occurs mainly in coastal localities in eastern, southern and south-west England and Wales, rarely inland; also recorded from southern Scotland and the west coast of Ireland. Found inland in Europe.

Other remarks: the species is most frequently collected during winter or spring, as suggested by the specific epithet.

Winter Stalk Puffball

Fig. 24. *Tulostoma brumale.* (Norfolk, Thornham, Dec. 1992, Outen)

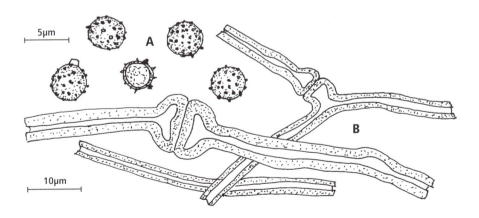

Fig. 25. *Tulostoma brumale.* (Sussex, Littlehampton, Dec. 1963, Rayner). A. Spores; B, paracapillitial threads.

3. SCALY STALK PUFFBALL

Map No. 11

Tulostoma melanocyclum Bres. in Petri in *Ann. Mycol.* 2: 415 (1904).

Selected descriptions: Dring (1966: 169–171); Wright (1987: 149–151)

Selected illustrations: Dring (1966: fig. 1, d–g); Mornand (1989: fig. 1); Pouzar (1958: figs 218, 219A); Wright (1987: fig. 95; pl. 26/ 1–2, 4; 30/ 5).

Diagnostic characters: Spores verruculose, 5–6.5 µm diam. (excluding ornament); septa of capillitial hyphae scarcely inflated, not or slightly pigmented; exoperidium hyphal, inconspicuous; socket lacking distinct lacerate membrane.

Fruitbody epigeous at maturity, stipitate, 25–40 mm high, base rooting, binding sand grains. *Stipe* rooting, cylindric or slightly tapered below, 23–35 (–50) mm high, 2–3 (–4) mm diam., brownish or reddish-brown, sometimes yellowish above, surface fibrous, often scaly towards the apex, binding sand grains at the base. *Exoperidium* hyphal, tending to bind sand grains. *Endoperidial body* 8–10 mm diam., subglobose, depressed beneath, with a torn collar-like zone around stipe apex, peristome single. *Endoperidium* cream to pale straw or pinkish, finely yellowish scurfy-pruinose. *Peristome* apical, well-defined, circular, slightly raised, somewhat darker than surrounding endoperidium. *Gleba* yellow-brown.

Capillitial hyphae 3–10 µm diam., hyaline, branched, thick-walled, with lumen continuous but often narrow, septate, septa not or only slightly inflated, hyaline or brownish, surface smooth or slightly encrusted. *Basidia* not seen. *Basidiospores* globose or subglobose, pale yellowish brown, with an irregular, minutely verruculose ornament, 5–6 (–6.7) µm diam excluding ornament, 5.5–7 (–7.5) µm diam. including ornament, verruculae isolated or rarely coalescent, 0.2–0.5 µm high.

Habitat: on sandy soil, mostly on dunes, amongst moss or herbaceous plants.

Distribution and frequency: very rare in Britain, known largely from western and south-western localities in Wales, Devon and Lancashire; there is also a single collection from Norfolk. It is widely distributed in Europe, particularly in southern parts, and is known also from North America. Records from Asia and South America require confirmation (Wright 1987).

Other remarks: First reported from Britain by Dring (1966) based on six collections in Kew previously determined as *T. brumale*, including two revised by J.E.Wright as *T. brumale* var. *bataviense* Wright (ined.). Some additional collections are now available. The species occurs usually from October to December, but may also appear in spring. It is similar to *T. brumale*, but distinguished by the paler, pruinose endoperidium, the form of the exoperidium and by characters of the spores and capillitial hyphae.

Scaly Stalk Puffball

Fig. 26. *Tulostoma melanocyclum.* (Germany, Sandhausen, 10 Nov. 1979, Winterhoff).

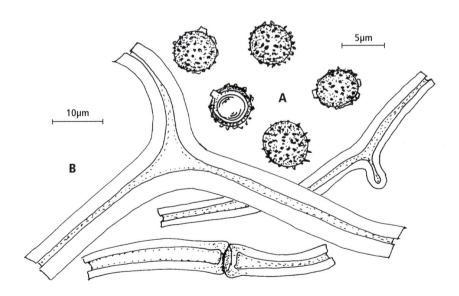

Fig. 27. *Tulostoma melanocyclum.* (Glamorgan, Porthcawl, Dec. 1970, Fitton). A, Spores; B, paracapillitial threads.

49

2. QUELETIA Fr.

Öfvers Forh. Kl. Svenska Vetensk.-Akad. 28: 171 (1871).

The genus is based on a single, rare, rather poorly known species and has remained monotypic until recently. A second species, *Q. andina* Wright (Wright 1989), described from Argentina, differs from *Q. mirabilis* most notably in having smooth spores. *Queletia* is similar in many respects to *Tulostoma*, and differs, primarily, only in its mode of dehiscence.

QUÉLET'S STALK PUFFBALL

Map No. 12

Queletia mirabilis Fr. in *Öfvers Forh. Kl. Svenska Vetensk.-Akad.* 28: 171, pl.4 (1871).

Selected descriptions: Azema (1990: 21); Coker & Couch (1928: 158–159); Mornand (1989: 13); White (1901: 441).

Selected illustrations: Cetto (1988: 380); Coker & Couch (1928: pl. 119 figs 16, 17); Dumée & Maire (1913: 501 figs 1–6; pl. XXVIII); Fischer (1900: fig. 180); Jülich (1983: pls 1–4 (spores)); Moravec (1958a: fig. 227); Mornand (1989: fig. 17); Ramsbottom (1953: pl. XII a, b); White (1901: pl. 38 figs 1–5).

Diagnostic characters: stipe well-developed, fibrous-scaly; peristome absent; spores coarsely verruculose, 6–7.5 μm diam. (excluding ornament).

Fruitbody at first sessile, subglobose or depressed, narrowed below, with sterile base, stipitate at maturity. *Stipe* 2.5–4.5 (–7) cm high, 0.5–1.5 (–2) cm diam., cylindric or tapered to apex, inserted into base of endoperidial body, dry, fibrous, surface splitting, scaly, whitish to pale brown. *Exoperidium* inconspicuous. *Endoperidial body* 2.5–4 (–5) cm diam., subglobose-depressed, somewhat ribbed below, irregularly lacerate or flared around stipe apex, dehiscing by irregular splitting and flaking of the endoperidium and easily separated from the stipe. *Endoperidium* scurfy or smooth, whitish to pale cream. *Gleba* pale brown, powdery at maturity.

   Capillitial hyphae mostly 5–9 μm diam., hyaline, simple or sparingly branched, undulate or irregularly coiled, tapered to rounded ends, 3–4.5 μm diam., thick-walled, with lumen distinct, sparsely septate, septa constricted, not pigmented, surface smooth or granulate. *Basidiospores* 5.6–7.5 μm diam. excl. orn., (7–) 7.5–9.5 μm diam. incl. orn., verruculae irregular, narowly conical or blunt, isolated or coalescent in small groups, 0.5–1.2 μm high, globose or subglobose, yellow-brown

Habitat: on old tanbark.

Distribution and frequency: an alien species, known in Britain from only two collections (a).October 1893 from the Herbarium grounds, Kew, found on discarded debris of Pennsylvanian material; (b) from spent tan at Barnsbury, London, October 1941.

Other remarks: this rare but distinctive species is known so far from only about ten collections, mostly from Europe and all on spent tan-bark. Whether it is native to Europe is uncertain, and its true distribution and ecology have yet to be ascertained. A full account of the appearance of the species in Britain is provided by Ramsbottom (1953).

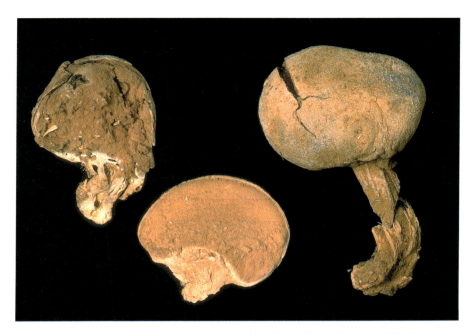

Fig. 28. *Queletia mirabilis*. Sections and surface view (Kew, Royal Botanic Gardens, 1893, Massee).

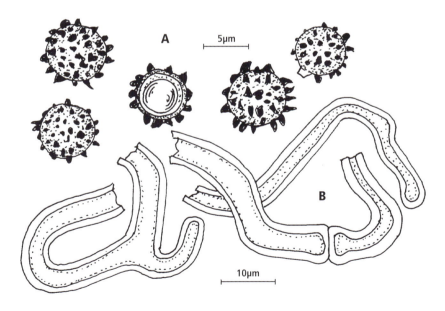

Fig. 29. *Queletia mirabilis*. (Kew, Royal Botanic Gardens, Oct. 1893, Massee). A. Spores; B, paracapillitial threads.

BATTARRAEACEAE Corda
in *Anleit. Stud. Mycol.* : 118 (1842, as *Battarreae*).

Fruitbody epigeous at maturity, stipitate, volvate, peridium 2-layered. *Exoperidium* gelatinous or membranous, at first egg-like, splitting to form volva and peridial cap. *Endoperidium* membranous, with circumscissile dehiscence or with several ostioles. *Stipe* cylindric, firm, woody, fibrous-scaly. *Gleba* brown, powdery at maturity.

Pseudocapillitium of broad, simple or rarely branched hyphae and elaters. *Elaters* cylindric, with conspicuous spiral or annulate thickenings. *Basidia* in clusters, not forming a hymenium, with 1–4 long, apical sterigmata. *Basidiospores* globose or subglobose, brown or yellowish, with verruculose or reticulate ornament. *Type genus*: *Battarraea* Pers.: Pers.

The family includes two genera, *Battarraea* and *Battarraeoides* Herrera (Herrera, 1953). The latter includes two Mexican species and differs from *Battarraea* principally in the ostiolate rather than cirmscissile dehiscence of the endoperidium. These genera have been commonly referred to *Tulostomataceae*. Their characters seem sufficiently distinct, however, to merit placement in a separate family. *Battarraeaceae* is accepted here in agreement with Maublanc & Malençon (1930), Jülich (1981), Perreau (1986) and Kreisel (1987).

STILT PUFFBALLS

BATTARRAEA Pers.: Pers.,
Synops. Meth. Fung. : 129 (1801, as *Batarrea*).

Dendromyces Libosch., *Beschr. Pilz.* (1814).

Exoperidium gelatinous or membranous. *Endoperidium* with circumscissile dehiscence. *Type species*: *Battarraea phalloides* (Dicks.: Pers.) Pers.

The orthography of the generic name, proposed in honour of the Italian mycologist G. A. Battarra, has received no consensus since the mis-spelling by Persoon (1801) as *Batarrea*. At least four other orthographic variants exist and the problem is discussed by Coetzee & Eicker (1992). The most logical spelling would seem to be as *Battarraea*, as proposed by Rauschert (1986) and accepted by various authors including those listed by Coetzee & Eicker and others such as Kubicka (1984), Mornand (1989), Pacioni (1983), Palmer (1968), Ramsbottom (1953), and Richter & Müller (1983).

SANDY STILT PUFFBALL

Map No. 13

Battarraea phalloides (Dicks.: Pers.) Pers., *Synops. Meth. Fung.*: 129 (1801).
Lycoperdon phalloides Dicks., *Fasc. pl. crypt. brit.* 1: 24 (1785).

Selected descriptions: Cooke (1871: 367); Lefevre (1982: 7–8); Pacioni (1983: 93); Perreau (1986: 1–3); Richter & Müller (1983: 61–63).

Selected illustrations: Bon (1987: p. 301); Cetto (1984: pl. 1637); Cetto (1988: 378); Cooke (1871: fig. 111); Lefevre (1982: 8–9); Moravec ((1958b: fig. 231); Mornand (1989: fig. 15); Pacioni (1983: 96); Perreau (1986: pl. 245); Phillips (1981: 250); Ramsbottom (1953: pl. 39b); Richter & Müller (1983: abb. 1).

Diagnostic characters: volva present, gelatinous; stem tall, hard, scaly; receptacle pendant; spirally thickened elaters.

Fruitbody egg 3–4 cm diam., ovoid, whitish at first, becoming brown, buried in soil; peridium 2-layered. *Exoperidium* gelatinous. *Endoperidium* with circumscissile dehiscence. *Stipe* at first with a gelatinous coat, soon dry, becoming hollow, 9–30 (–37) cm high, 6–20 mm diam., pale brown to brown or greyish brown, surface fibrous-scaly, often shaggy. *Receptacle* pendant, convex to hemispherical, bearing the gleba. *Gleba* exposed by shedding of peridial cap.
   Pseudocapillitium of mostly thin-walled, hyaline hyphae and elaters. *Elaters* hyaline, cylindric, tapered, sometimes branched, walls with conspicuous, refractive, spiral thickenings. *Basidia* not seen. *Basidiospores* (4.5–) 5–6 (–6.5) μm diam. excluding ornament, 5–6.5 (–7) μm diam. including ornament, yellowish or yellow-brown, densely verruculose, verruculae commonly coalescing to form anastomosing ridges.

Habitat: on dry, usually sandy soil in various places, sometimes associated with ash (*Fraxinus*), yew (*Taxus*) or pine (*Pinus*)

Distribution and frequency: very rare, known mainly from southern and eastern England. It was first collected, though in poor condition, by W. Humphrey at Bungay, Suffolk in 1782. Further material was collected by Woodward early in 1783 and published the following year (Woodward 1784). During the succeeding century this species was collected elsewhere in Suffolk and from a few localities in Norfolk, Cheshire, Surrey, Bucks and Kent. It was collected near Gloucester in 1915, and British collections to that date are summarized by Ramsbottom (1916 a, b; 1953). Since then, a few further collections have been made from Suffolk, Surrey, Kent, Avon, and Jersey. The most recent collection, in September 1975, was from the same area as the type gathering at Bungay, Suffolk. It is scarce in Europe. The species is reported also from North America, Africa and Australia, but records often include *B. stevenii* (Libosch) Fr. and are difficult to assess.

Other remarks: *B. stevenii* (Libosch.) Fr. , commonly treated as a distinct species, is considered synonymous with *B. phalloides* by some authors, notably Perreau (1986) and Kreisel (1987). It can be distinguished by its larger fruitbodies, up to 50 cm high, with thick, dry exoperidium. It also appears to differ in having a more southerly distribution, occuring in mediterranean and subtropical areas (see Mornand 1989). The situation in North America is uncertain. Both slender and robust forms were reported from California by Rea (1942), though the presence of a gel was not ascertained. Perreau (1986) suggests that this character may vary according to climatic conditions, and notes that no specific distinction can be made on microscopic characters. If *B. stevenii* proves to be conspecific with *B. phalloides*, it will provide an earlier name for the species.

Sandy Stilt Puffball

Fig. 30. *Battarraea phalloides* (Suffolk, Blyford, 7 Sept. 1984, Brand).

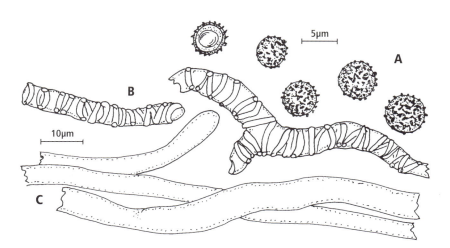

Fig. 31. *Battarraea phalloides*. (Kent, Wicham, June 1955, Martin). A. Spores; B, elaters; C, pseudocapillitial threads.

BIRD'S NEST FUNGI & CANNON FUNGUS

NIDULARIALES G. Cunn.
Gasterom. Austr. & New Zeal. : 198 (1944).

Fruitbody epigeous, terrestrial on rotten vegetation or lignicolous, sometimes on dung, small, 1–10 mm diam., globose to cupulate, sessile, often gregarious. *Peridium* of 1–3 layers, with apical dehiscence. *Gleba* of one to many, separate peridioles. *Capillitium* absent. *Hymenium* not developed; *basidia* borne singly or in groups, 4–8-spored. *Basidiospores* often large, broadly ellipsoid to cylindrical, hyaline, smooth-walled. *Type family*: *Nidulariaceae* Dumort.

Key to Families

1. Fruitbody 1–10 mm diam., globose, short cylindrical to obconical, dehiscing either . irregularly by apical fragmentation of the peridium or by a circumscissile epiphragm; usually several, ellipsoid peridioles; peridiole release passive, by water-splash action .... **1. Nidulariaceae**
1. Fruitbody up to 3 mm diam.; globose, with stellate, apical dehiscence of the peridium to reveal a single, globose peridiole; inner peridial wall everts to allow violent release of peridiole .... **2. Sphaerobolaceae**

1. NIDULARIACEAE Dumort.
in *Comment. Bot.* 69 : 81 (1822).

Description as for order but peridiole release passive, by water-splash action. *Type genus*: *Nidularia* Fr. & Nordh.

Key to British Genera

1. Peridiole attached to the inner peridial surface by a thread-like funiculus, not immersed in mucilage; fruitbody cupulate, mostly more than 5 mm diam. 2
1. Peridioles free within the peridium, immersed in mucilage; fruitbody subglobose, generally less than 5 mm diam. 3

    2. Peridium consisting of more than one layer, with grooved or smooth inner surface; peridioles pale grey to black .... **1. Bird's Nest Fungi** (*Cyathus*)
    2. Peridium consisting of one layer; fruitbody short cylindrical, with smooth inner surface; peridioles whitish .... **2. *Crucibulum***

3. Fruitbody 5–7 mm diam.; peridium thick, cream-coloured to cinnamon, splitting apically to reveal numerous, dark peridioles; hyphal system dimitic, consisting of pale brown, spinose, non-septate tapered hyphae plus thread-like and branched, hyaline, thin-walled generative hyphae with clamp-connexions .... **3. *Nidularia***
3. Fruitbody up to 2 mm diam., often much less; peridium thin, whitish or nearly so, either irregularly indehiscent or evanescent; hyphal system monomitic, of thin-walled, hyaline generative hyphae with clamp-connexions .... **4. *Mycocalia***

1. BIRD'S NEST FUNGI

CYATHUS Haller : Pers.,
Synops. Meth. Fung. : 237 (1801); Haller, *Hist. Stirp. Helvet.* 3 : 127 (1768).

Fruitbody epigeous, cupulate or infundibuliform at maturity with a broad apical rim and a narrow base, initially covered by an epiphragm. *Peridium* stratified, outer surface tomentose, shaggy or squamose, internal surface glabrous, smooth or ridged. *Epiphragm* thin, membranous, pale coloured, circumscissile, ephemeral. *Peridioles* numerous, generally more than 10, dark coloured, thick-walled and sometimes with an outer, hyaline tunica; each peridiole is attached at a depression on the side to the peridial wall by a long, complex funiculus.

Basidiospores more than 10 µm long, hyaline, smooth, with a thickened wall. *Peridiopellis* three-layered, outer layer of densely woven, narrow hyphae running parallel to surface, middle layer pseudoparenchymatous, inner layer of loosely woven hyphae. *Type species*: *Cyathus olla* Batsch: Pers.

Cyathus is the largest of the genera within the Bird's Nest Fungi, with a worldwide distribution although only three species are reported from Britain and Europe. The gasterocarps are most frequently found in tropical-subtropical localities, where about 40 species are known. Their appearance is sporadic and they cannot always be demonstrated. When they do occur, however, they may be found in enormous numbers, generally growing in clusters on the ground, on fallen wood and debris, decaying herbaceous material, leaf litter and dung.

Key to British Species

1. Peridium with the inner surface conspicuously striate to fluted, greyish brown, and outer surface hairy, dark brown; fruitbody infundibuliform with vertical or slightly flared margin; spores 12–21 x 7–12 µm, oblong ellipsoid; on the ground or on plant remains .... **1. Fluted Bird's Nest** (*C. striatus*)
1. Peridium with the inner surface smooth, greyish to blackish, lacking striations or ridges 2

    2. Spores very large, 23–33 x 17–30 µm, subglobose; fruitbody cupulate to broadly infundibuliform with a vertical margin; peridium hairy, ochre-brown to blackish brown; peridioles 1–2 mm diam., black; on dung or manured soil
    .... **2. Dung Bird's Nest** (*C. stercoreus*)
    2. Spores 9–12 x 6.5–8.5 µm, broadly ellipsoid; fruitbody infundibuliform with a wide, flaring margin; peridium greyish yellow to brown; peridioles 2–3 mm diam., grey-brown; on humus, more rarely on plant remains
    .... **3. Field Bird's Nest** (*C. olla*)

1. FLUTED BIRD'S NEST

Map No. 14

Cyathus striatus (Huds.) Pers., *Synops. Meth. Fung.*: 237 (1801).
Peziza striatus Huds., *Fl. Angl.* Edit.2, 2: 634 (1778).
Nidularia striata (Huds.) With., *Bot. Arrang.* Edit. 2, 3: 446 (1792).
Nidularia hirsuta (Schaeff.) Sow., *Engl. Fung.* 1: pl. 29 (1796).

Selected descriptions: Brodie (1975: 173); Cejp (1958: 656, figs. 244–246); Marchand (1976: 154, pl.375); Massee (1889: 54, fig. 48); Rea (1922: 46); Wakefield & Dennis (1981: 209, pl.111/9).

Selected illustrations: Brand (1988: 110); Breitenbach & Kränzlin (1986: pl. 496); Gerhardt (1985: 183); Jahn (1979: pl. 210); Pegler (1990: 168); Phillips (1980: 254); Ramsbottom (1953: pl. 14d); Ryman & Holmåsen (1984: 583); Sowerby (1796: pl. 29).

Diagnostic characters: infundibuliform fruitbody with brown, shaggy surface; inner surface strongly striated or fluted; spores 12–21 x 7–12 µm.

Fruitbody 8–15 mm tall, 6–8 mm wide, strongly infundibuliform with a narrow, tapering base. *Peridium* at first entirely covering the gasterocarp then apically fragmenting to reveal the epiphragm, outer surface rusty brown to dark fuscous brown, shaggy-tomentose to hairy; inner surface greyish, vertically ridged or fluted. *Peridioles* 12–16 in number, 1–2 mm diam., lenticular, pale greyish, each attached by a fine, thread-like funiculus to the inner peridial surface.
   Basidiospores 12–21 x 7–12 µm, oblong ellipsoid to ellipsoid, hyaline, smooth, thick-walled (–1.5 µm). *Peridiopellis* of brown, constricted hyphae, with clamp-connexions, and fusoid terminal elements, 35–75 x 8–15 µm.

Habitat: gregarious on fallen branches, twigs and other debris, often in large numbers, nearly always in woodland, rarely in gardens.

Distribution and frequency: common in England, less so in Scotland, fruiting especially in summer and autumn. Occurs throughout the temperate regions.

Other remarks: the species was originally described from England. Fruitbodies of *Crucibulum laeve* are similar in form but lack the peridial fluting, have much smaller spores, and spinose hyphae in the peridiopellis.

Fig. 32. *Cyathus striatus* (Surrey, Virginia Water, 20 Aug. 1992, Legon).

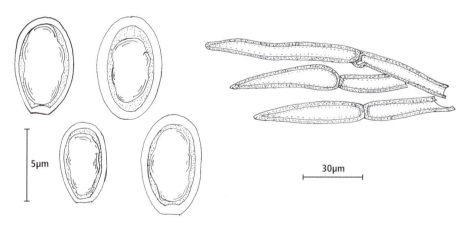

Fig. 33. *Cyathus striatus* (Bedfordshire, Clophill, 12 Nov. 1973, Outen). Spores and peridiopellis hyphae.

2. DUNG BIRD'S NEST

Map No. 15

Cyathus stercoreus (Schwein.) De Toni in Sacc., *Syll. Fung.* 7: 40 (1888).
Nidularia stercorea Schwein. in *Trans. Amer. Phil. Soc.* 4: 253 (1832).

Selected descriptions: Brodie (1975: 168, fig.59a); Brodie & Dennis (1954: 156); Cejp (1958, figs. 242–243).

Selected illustrations: Breitenbach & Kränzlin (1986: pl.495); Rotheroe, Hedger & Savidge (1987: 15).

Diagnostic characters: blackish, shaggy fruitbodies; smooth inner peridial surface; black peridioles; large, subglobose spores; coprophilous habitat.

Fruitbody small, 6–10 mm tall, 4–8 mm wide, barrel-shaped to cupulate, finally infundibuliform but not developing a flaring margin. *Peridium* initially enveloping entire fruitbody then rupturing at apex to expose underlying epiphragm, ochre-brown to blackish brown, initially adpressed shaggy-tomentose on outer surface but glabrescent; inner surface dark grey, smooth. *Peridioles* 1–2 mm diam., black, each attached by a fine, thread-like funiculus to the peridial wall.
    Basidiospores 23–33 x 17–30 µm, subglobose, hyaline, thick-walled (–5µm), smooth, with granular contents. *Peridiopellis* hyphae hyaline to pale yellowish, thick-walled, with the terminal elements lanceolate, 70–130 x 6–12 µm.

Habitat: in Britain only in sand dunes on rabbit pellets, sometimes attached to marram grass. Elsewhere on dung and manured soil, bonfire sites, and perhaps also lignicolous.

Distribution and frequency: very rare in Britain, known from dune systems in Wales and Galloway. Almost cosmopolitan.

Other remarks: in Britain rather uniform in appearance, but elsewhere variable in form and can resemble other other species of *Cyathus*. *Cyathus olla* can also occur in dunes. Brodie (1975) reports that *C. stercoreus* grows easily in culture and that its morphology can be considerably influenced by environmental factors.

Dung Bird's Nest

Fig. 34. *Cyathus stercoreus* (Dyfed, Ynyslas, Rotheroe).

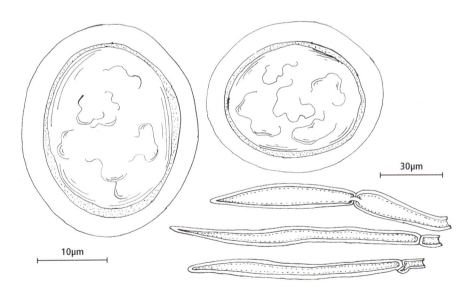

Fig. 35. *Cyathus stercoreus* (Dyfed, Ynyslas, 5 Oct. 1983, Hedger). Spores and peridiopellis hyphae.

3. FIELD BIRD'S NEST

Map No. 16

Cyathus olla (Batsch: Pers.) Pers., *Synops. Meth. Fung.*: 237 (1801); Batsch, *Elench. Fung.* 1: 127 (1763, as *Peziza*).
Nidularia campanulata With., *Bot. Arrang.* edit. 2, 3: 445 (1792).
Cyathus vernicosus DC, *Fl. Franç.* 2: 270 (1805).

Selected descriptions: Brodie (1975: 154); Cejp (1958: 648, figs. 240–241); Massee (1889: 55, figs. 49–51); Rea (1922: 47); Wakefield & Dennis (1981: 209, pl.111/10).

Selected illustrations: Breitenbach & Kränzlin (1986: pl.494); Gerhardt (1985: 184); Pegler (1990: 168); Phillips (1982: 254); Sowerby (1796: pl.28).

Diagnostic characters: fruitbody with wide, flaring margin, at first with silky outer surface; inner surface smooth, greyish; peridioles relatively large, grey; spores 9–12 x 6.5–8.5 µm.

Fruitbody initially egg-shaped, later strongly infundibuliform, 8–15 mm high and wide, developing a widely flaring margin. *Peridium* greyish yellow to greyish brown, with a silky-tomentose, soon glabrescent outer surface; smooth and silver-grey on inner surface. *Epiphragm* whitish, membranous, finally rupturing to expose peridioles. *Peridioles* 8–10 in number, relatively large, 2–3 mm diam., grey; each attached by a fine, thread-like funiculus.
  Basidiospores 9–12 (–14) x 6.5–8.5 µm, ovoid to broadly ellipsoid, at times subpyriform, hyaline, smooth, with a slightly thickened wall and granular contents. *Peridiopellis* with outermost layer of hyaline to yellowish brown hyphae, with thickened walls, clamp-connexions, and terminating in elongate, clavate to fusoid elements, 40–110 x 6.5–8.5 µm.

Habitat: on twigs, humus and fallen debris. Often found in gardens on compost, in plant pots, on old boards and wood chippings, and in dunes.

Distribution and frequency: widespread in Britain and the most common species in Europe, although not found in the most northerly regions. Also known from North & South America, South Africa, Australia, but not tropical.

Other remarks: easily recognized by the subglabrous surface, the flared peridial margin, and the large peridioles. The variety ***agrestis*** (Pers.) Tul. differs in the smaller, subglobose fruitbody, with a non-flaring peridial margin. The forma ***anglicus*** (Lloyd) Brodie was described from England and distinguished by the particularly large size, up to 15–18 mm, with peridioles up to 4 mm diam. *Crucibulum laeve* can be confused but the one-layered peridiopellis contains brown, spinose hyphae.

Field Bird's Nest

Fig. 36. *Cyathus olla* (Devon, Dawlish Warren, 19 Oct. 1985, Roberts).

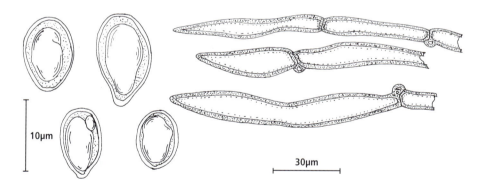

Fig. 37. *Cyathus olla* (Surrey, Kew, Royal Botanic Gardens, 26 June 1990, Laessøe). Spores and peridiopellis hyphae.

2. CRUCIBULUM Tul. & C. Tul.
in *Ann. Sci. Nat., Bot.* sér. 3, 1 : 89 (1844).

Only species: *Crucibulum laeve* (Huds.) Kambly.

WHITE-EGG BIRD'S NEST

Map No. 17

Crucibulum laeve (Huds.) Kambly in Kambly & Lee, *Gasterom. Iowa*: 167 (1936, ut *levis*).
Nidularia laevis Huds. in Relhan, *Fl. Cantabr.* Edit. 2: 529 (1802); Huds, *Fl. Angl.* Edit. 2, 2: 634 (1778 ut *Peziza laevis*).
Cyathus crucibulum Pers., *Synops. Meth. Fung.*: 238 (1801).
Crucibulum vulgare Tul. & C. Tul. in *Ann. Sci. Nat., Bot.* sér. 3, 1: 90 (1844).

Selected descriptions: Brodie (1975: 148); Cejp (1957: 640, figs. 236–239); Rea (1922: 46); Tulasne & C. Tulasne (1844: 90, pl.6/9–24, 7/1, 18–25; 8/13–17); Wakefield & Dennis (1981: 208, pl.111/7).

Selected illustrations: Breitenbach & Kränzlin (1986: pl.493); Gerhardt (1985: 185); Jahn (1979: pl.209); Massee (1889: fig. 52); Pegler (1990: 168); Phillips (1982: 254); Ryman & Holmåsen (1984: 584).

Diagnostic characters: fruitbody of short-cylindrical cups, initially with ochraceous epiphragm; peridium single-layered, inner surface pale and smooth; peridioles numerous, whitish, each with funiculus.

Fruitbodies gregarious, mostly 5–8 mm high., sometimes up to 10 mm high, 7–8 mm diam., almost globose then cup-like to short-cylindrical, sessile, soft to firm, with peridium apex initially covered by a fugacious, membranous epiphragm, with the margin finally expanding after exposure. *Peridium* thick, felt-like, greyish to cinnamon-brown, finally blackish brown, minutely tomentose, glabrescent, internally smooth, creamy white and shiny, comprising only one layer of more or less thick-walled, brown hyphae, 2–3 µm diam., becoming spinose in the tomentum. *Epiphragm* operculate, yellow to ochraceous, rupturing at maturity. *Peridioles* numerous, 10–15 (–20), not embedded in mucilage, 1.5–2 mm diam., lenticular, pale ochraceous then whitish, each attached to peridial wall by a filiform funiculus.
   Basidiospores 6–10 x 3.5–5 µm, oblong ellipsoid to ellipsoid, hyaline, smooth, with a slightly thickened wall. *Surface hyphae of peridium* 3–7 µm diam., thick-walled, bright yellowish brown, with numerous, irregular spinose outgrowths.

Habitat: growing on wood, twigs, decomposing herbaceous stems and compost.

Distribution and frequency: widespread throughout the British Isles, and fairly common in autumn and winter. Worldwide in distribution.

Other remarks: Frequently listed as *C. vulgare*. This and *Cyathus striatus* are the most common species of the bird's nest fungi.

White-Egg Bird's Nest

Fig. 38. *Crucibulum laeve*. (July 1992, Laessøe).

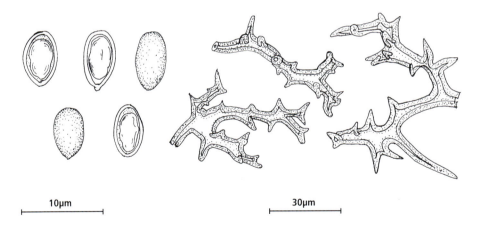

Fig. 39. *Crucibulum laeve* (Glamorgan, Merthyr Mawr, 15 Sept. 1973, Ing). Spores and peridiopellis hyphae.

3. NIDULARIA Fr. & Nordh.,
Symb. Gast. : 2 (1817).

Only species: *Nidularia deformis* (Willd.: Pers.) Fr.

PEA-SHAPED BIRD'S NEST

Map No. 18

Nidularia deformis (Willd.: Pers.) Fr. in Fr.& Nordh., *Symb. Gast.*: 3 (1817).
Cyathus deformis Willd: Pers., *Syn. Fung.*: 246 (1801); Willd. apud Roem. & Usteri in *Bot. Mag.* 2(4): 14 (1788).
Nidularia confluens Fr. in Fr. & Nordh., *Symb. Gast.*: 3 (1817).
Nidularia farcta (Roth.: Pers.) Fr., *Syst. Mycol.* 2: 301 (1823).
Nidularia pisiformis Tul. & C.Tul. in *Ann. Sci. Nat., Bot.* sér. 3, 1: 100 (1844).
Nidularia pisiformis var. *broomei* Massee in *Ann. Bot., Lond.* 4: 58 (1889).
Nidularia berkeleyi Massee in *Ann. Bot., Lond.* 4: 59 (1889).

Selected descriptions: Breitenbach & Kränzlin (1986: 380); Cejp (1958: 661–665, figs. 247–249); Jahn (1972); Massee (1889: 58, figs. 37–38); Rea (1922: 45).

Selected illustrations: Breitenbach & Kränzlin (1986: pl.497); Buczacki (1989: 197); Currey (1863: 151, pl.25/ 4–6 & 21–22); Gerhardt (1985: 182); Phillips (1981: 254); Ramsbottom (1953: pl.14b); Ryman & Holmåsen (1984: 584); Wakefield & Dennis (1981: pl.111/8).

Diagnostic characters: fruitbodies to 1 cm diam., clustered on dead wood; peridioles not attached to inner peridial wall; peridium tearing irregularly over the apex, cream-coloured; hyphal system dimitic, with pale brown, spinose hyphae.

Fruitbody small, 0.3–1 cm diam., globose to pea-shaped, finally tuberculate; sessile and broadly attached at the base. *Peridium* creamy white to greyish or cinnamon brown, finely woolly-floccose, consisting of a single layer, internally smooth, at first completely enclosing the peridioles then irregularly tearing over the apex, upper portion finally falling away. *Peridioles* numerous, 1–2 mm diam., lenticular, chestnut-brown, smooth and shiny, not attached by a funiculus, immersed in mucilage.
Basidiospores 5–9.5 x 4–6 μm, ovoid to broadly ellipsoid, hyaline, smooth, with a slightly thickened wall. *Peridial hyphae* 3–6 μm diam., with numerous short branches and spinose outgrowths, hyaline to pale brown, with a slightly thickened wall.

Habitat: gregarious on dead, fallen branches or wood-chips.

Distribution and frequency: Widespread but uncommon in Britain, fruiting during summer and autumn. Restricted to northern Europe, where it is the only species of *Nidularia* Fr.

Other remarks: more widely known under the name *N. farcta*. Two varieties have been proposed: var. *radicata* Fr. with conspicuous mycelium, and var. *confluens* (Fr.) Cejp with confluent fruitbodies.

Pea-shaped Bird's Nest

Fig. 40. *Nidularia deformis.* (Denmark, Raunshalt, 2 Nov. 1985, Laessøe).

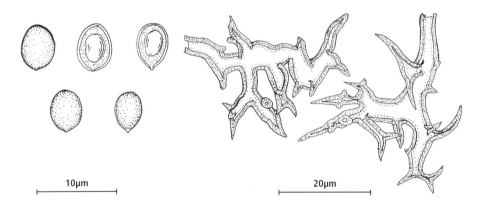

Fig. 41. *Nidularia deformis* (Warwickshire, Water Orton, 23 Aug. 1980, Clark 2241). Spores and peridiopellis hyphae.

MYCOCALIA Palmer
in *Taxon* 10: 58 (1961).

Nidularia Fr. & Nordh., *Symb. Gast.* 1: 2 (1817), pro parte
Granularia Roth ex Nees, *Hor. Phys. Berol. Collect.*: 6 (1820), pro parte.
Nidularia sect. *Sorosia* Tul. & C. Tul. in *Ann. Sci. Nat., Bot.* sér. 3, 1: 98 (1844).

Fruitbody epigeous, gregarious or solitary, minute, 0.25–2 mm diam., more or less globose, at times confluent. *Peridium* thin, flocculose, evanescent, whitish or nearly so, non-dehiscent or irregularly so at the apex. *Hyphal system* monomitic; generative hyphae hyaline, thin-walled, with clamp-connexions. *Peridioles* one to several, lenticular, not attached by a funiculus to peridial wall.

Basidiospores subglobose to subfusoid, hyaline to pale brown, smooth, with a slightly thickened wall. *Basidia* metamorphosed, often thick-walled and coloured. *Type species: Nidularia denudata* Fr. [= *Mycocalia denudata* (Fr.) Palmer.]

Key to British Species

1. Spores 9–17 x 4.5–5.5 µm, pale brown, subfusoid; peridium of one layer only, containing a solitary peridiole, 0.5 mm diam. .... **1. Bog Mycocalia** (*M. sphagneti*)
1. Spores up to 8.5 µm long, hyaline, ovoid; peridium two-layered, containing one to many peridioles 2

    2. Peridiole always solitary, c. 220 µm diam.; metamorphosed basidia ovoid to globose with a truncated base; spores 4–6.5 x 3–4 µm
    .... **2. Tiny Mycocalia** (*M. minutissima*)
    2. Peridioles one to many, each more than 250 µm diam.; metamorphosed basidia ellipsoid to piriform; spores 5.5–8.5 x 4.5–7 µm 3

3. Peridioles yellowish brown, appearing collapsed on drying; outer cortical layer loosely woven .... **3. Common Mycocalia** (*M. denudata*)
3. Peridioles dark red to almost black, not readily collapsing when dry; outer cortical layer of dense, compacted hyphae .... **4. Durieu's Mycocalia** (*M. duriaeana*)

Species of the genus *Mycocalia* were formerly included in *Nidularia*, until Palmer (1961) separated the former on the basis of the ephemeral, almost non-existent peridium consisting only of clamp-bearing generative hyphae. Several species of *Mycocalia* are distinguished mainly by their habitat. They usually grow on plant materials in wet areas, especially on *Juncus* species, although *M. duriaeana* is adapted to arid, sandy locations. Both *M. denudata* and *M. minutissima* are probably very common and widespread but often overlooked.

1. BOG MYCOCALIA

Map No. 19

Mycocalia sphagneti Palmer apud Cejp & Palmer in *Ceská Mykol.* 17: 122 (1963).

Selected descriptions: Brodie (1975: 140); Cejp & Palmer (1963: 122).

Selected illustrations: Cejp & Palmer (1963: pl.13/2, 14/4-5; fig. 6-7).

Diagnostic characters: peridium one-layered; solitary peridiole, 0.5 mm diam.; spores 9–17 x 4.5–6.5 µm , subfusoid, pale brown.

Fruitbodies solitary or gregarious, but never confluent. *Peridium* white, at first woolly, dry, later smooth, closely adherent to the peridiole, arachnoid and finally evanescent. *Peridiole* solitary, variable in size, 450–685 µm diam., and 100–300 µm thick, initially white, soon dark blood-red to black, darker towards the centre, soon collapsing, smooth, with a single cortical layer of compacted hyphae.

Basidiospores 9–17 x 4.5–6.5 µm, elongate ellipsoid to subfusoid, hyaline or pale yellowish brown, smooth, with a distinctly thickened wall. *Basidia* 25 x 8 µm, clavate, hyaline, 2–4 sterigmate; *metamorphosed basidia* 12 x 10 µm, ellipsoid to pyriform, usually basally truncated.

Habitat: on *Juncus* culms in *Sphagnum*-bogs.

Distribution and frequency: described from Derbyshire; and now known also from Hampshire. Elsewhere in Europe it has been reported only from Sweden.

Other remarks: the one-layered cortex of the peridiole, and the large, yellowish brown spores are distinctive.

2. TINY MYCOCALIA

Map No. 20

Mycocalia minutissima (Palmer) Palmer in *Taxon* 10: 58 (1961).
Nidularia minutissima Palmer in *Naturalist (Lond.)* 1957: 4 (1957).

Selected descriptions : Brodie (1975: 138); Cejp & Palmer (1963: 120); Palmer (1963: 16).

Selected illustrations: Cejp (1958: fig. 252); Cejp & Palmer (1963: pl. 13/5, 14/2; fig. 4); Ellis & Ellis (1990: fig. 523); Palmer (1963: figs 5-10).

Diagnostic characters: peridium of two layers; spores 4–6.5 x 3–4 µm, ovoid, hyaline; peridiole solitary, about 220 µm diam.; basidia ovoid, basally truncated.

Fruitbodies solitary or gregarious but not confluent, minute, up to 250 µm diam., subglobose. *Peridium* thin, loose, reticulate-scurfy, white, appearing moist, finally irregularly rupturing or evanescent to reveal a solitary peridiole. *Peridiole* solitary, embedded in gelatinous matrix, 150–230 µm diam., lenticular, collapsing when dry, deep yellow to dark brick-red, smooth, with a two-layered cortex.
   Basidiospores 4–6.5 x 3–4 µm, ovo-ellipsoid, hyaline, smooth, with a thickened wall. *Metamorphosed basidia* 7–11.5 x 6–9 µm, globose to irregular, hyaline.

Habitat: always in moist situations, on *Juncus* and grass culms, birch (*Betula)* leaves, moss, and decorticated *Pinus* wood.

Distribution and frequency: originally described from Lancashire; widespread and probably quite common but difficult to collect. Otherwise known from former Czechoslovakia, Germany and Sweden; mostly at higher altitudes (Palmer, 1958).

Other remarks: The fruitbody is reduced to one peridiole surrounded by an evanescent peridial layer. Scattered peridioles and developing gasterocarps sometimes found in submerged situations. Closely related to *M. denudata* and *M. duriaeana* and difficult to distinguish from uniperidiolar fruitbodies of those species.

3. COMMON MYCOCALIA

Map No. 21

Mycocalia denudata (Fr.) Palmer in *Taxon* 10: 58 (1961).
Nidularia denudata Fr. apud Fr. & Nordh., *Symb. Gast.*: 2 (1817).
Cyathus denudatus (Fr.) Spreng. apud Linn, *Syst. Veg.* Edit. 16, 4 (1): 415 (1827).
Granularia denudata (Fr.) Kuntze, *Rev. Gen. Pl.* 2: 855 (1891).
Nidularia arundinacea Velen., *Nov. Myc.*: 169 (1939).
Nidularia fusispora Massee in *Bull. Misc. Inf. Kew* 1898: 125 (1898).

Selected descriptions: Brodie (1975: 138); Buczacki (1989 : 197); Cejp (1958: 666–669); Cejp & Palmer (1963: 117); Palmer (1963: 14); Calonge & Palmer (1988).

Selected illustrations: Buczacki (1989 : 197); Cejp (1958: fig. 250); Cejp & Palmer (1963: pl. 13/4, 14/1; figs 1–2); Elborne (1983: 45); Ellis & Ellis (1990: fig. 521); Palmer (1963: figs 1–4).

Diagnostic characters: small size; peridium of two layers; spores 5.5–8.5 x 4.5–7 µm, hyaline, ovoid; peridioles one to many, yellowish brown, collapsed on drying; outer cortical layer loosely woven.

Fruitbody 0.2–1.5 mm diam., subglobose but often confluent. *Peridium* delicate, finally rupturing and evanescent, white or slightly yellowish, thin, flocculose; exoperidium of loosely woven hyphae, often exuding water droplets. *Peridioles* numerous, 300–400 µm diam., 70–140 µm thick, lenticular-discoid when fresh becoming biconcave on drying, free but embedded in a hyaline, gelatinous matrix; outer cortical layer yellowish, of loosely woven hyphae; inner cortical layer thicker, reddish brown.
  Basidiospores 5.5–8.5 x 4.5–7 µm, broadly ovoid, hyaline, inamyloid, with a thickened wall. *Metamorphosed basidia* 9–12 x 7–9 µm, globose, ellipsoid to pyriform, thick-walled.

Habitat: either on dead wood in wet places or on stems of *Juncaceae* and *Cyperaceae*, throughout the year, preferring acid conditions.

Distribution and frequency: Common and widespread in Britain, occurring particularly at higher altitudes (Palmer, 1958). Widespread throughout northern Europe, and also reported from Australia, North America (British Columbia), Greenland and South America.

Other remarks: The most commonly recorded species of *Mycocalia*.

4. DURIEU'S MYCOCALIA

Map No. 22

Mycocalia duriaeana (Tul. & C. Tul.) Palmer in *Taxon* 10: 58 (1961).
Nidularia duriaeana Tul.& C. Tul. in *Ann. Sci. Nat., Bot.* sér. 3, 1: 99 (1844).
Granularia duriaeana (Tul. & C. Tul.) Kuntze, *Rev. Gen. Pl.* 2: 855 (1891).
Nidularia castanea (White) P. & D. Sacc. in P.Sacc., *Syll. Fung.* 17: 216 (1905).

Selected descriptions: Brodie (1975: 138); Cejp & Palmer (1963: 119); Tulasne & C. Tulasne (1844: 99–100).

Selected illustrations: Cejp (1958: fig.251); Cejp & Palmer (1963: pl.14/ 3 and 6; fig. 3); Ellis & Ellis (1990: fig. 522); Tulasne & C. Tulasne (1844: pl.7/13–17).

Diagnostic characters: peridium of two layers; peridioles one to many, dark reddish to black, not collapsing; outer cortical layer densely compacted; spores 5.5–7.5 x 4–5.5 µm, hyaline, ovoid; basidia ellipsoid to pyriform.

Fruitbody up to 1 mm diam., globose. *Peridium* very thin, white, evanescent. *Peridioles* about 300 µm diam., and 150 µm thick, dark reddish brown to almost black, lenticular and not collapsing when dry; with a thick, double cortical layer of compacted hyphae.
   Basidiospores 5.5–7.5 x 4–5.5 µm, ovoid, hyaline, smooth, with a thickened wall. *Basidia* subsessile, ovoid to pyriform.

Habitat: known from sand-dunes, attached to culms of *Ammophila*, also on rabbit dung, dead twigs and pine needles, usually formed on upper surface of substratum, along the line of contact with the sand.

Distribution and frequency: Rare in the British Isles, known only from Lancashire. Elsewhere, known from north Africa and Europe, the east coast of North America, and from Tasmania.

Other remarks: Originally described from Algeria, and named after the collector and botanist, Durieu de Maisonneuve. Very similar to *M. denudata* but differing in the very dark peridioles which have a compacted outer cortex, and do not collapse on drying. The species may be quite common but the peridioles closely resemble the adjacent sand-grains making recognition in the field difficult.

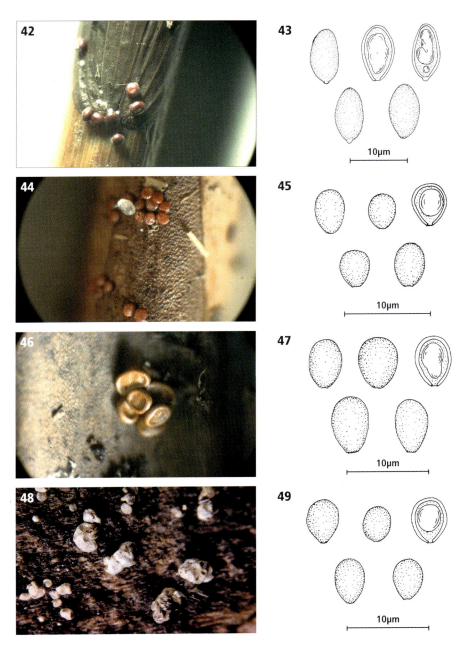

Fig. 42. *Mycocalia sphagneti*. (Derbyshire, Bleeklaw, Palmer). Fig. 43. *Mycocalia sphagneti* (Derbyshire, Hayfield, 23 March 1963, Palmer 11473). Spores. Fig. 44. *Mycocalia minutissima* (On *Juncus*, Palmer). Fig. 45. *Mycocalia minutissima* (Derbyshire, Hoo Moor, April 1956, Palmer). Spores. Fig. 46. *Mycocalia denudata* (Derbyshire, Palmer). Fig. 47. *Mycocalia denudata* (Norfolk, Buxton Heath, 5 Oct. 1958, Palmer). Spores. Fig. 48. *Mycocalia duriaeana* (USA, New Jersey, 26 Oct. 1961, Palmer). Fig. 49. *Mycocalia duriaeana* (Lancashire, Formby, 1 Sept. 1956, Palmer). Spores.

2. SPHAEROBOLACEAE J. Schröt.
apud Cohn, *Kryptog.-Fl. Schlesien* 3 (1) : 688 (1889).

Fruitbody gregarious, subglobose to ovoid, sessile. *Peridium* multilayered, including gelatinized layers, splitting apically with stellate rays at maturity. *Peridiole* solitary, globose, dark, containing both basidiospores and chlamydosporic gemmae; forcibly ejected by eversion of the inner peridial layer which separates from the pseudo-parenchymatous layer. *Spore mass* white.

Basidiospores small, oblong to ellipsoid, hyaline, smooth. *Basidia* clavate, bearing numerous spores. On rotting plant debris and dung. *Type and only genus*: *Sphaerobolus* Tode: Pers.

CANNON FUNGUS

SPHAEROBOLUS Tode : Pers.,
Synops. Meth. Fung. : 115 (1801).

Only species: *Sphaerobolus stellatus* Tode: Pers.

Map No. 23

Sphaerobolus stellatus Tode : Pers., *Synops. Meth. Fung.*: 115 (1801); Tode, *Fung. Mecklenb.* 1 : 43 (1790).
Lycoperdon carpobolus L., *Spec. Plant.* 2: 1654 (1753).
Carpobolus stellatus (Tode : Pers.) Mich. ex Desm., *Obs. bot.- zool.*: 9 (1826).
Sphaerobolus carpobolus (L.) Schröt., *Pilz. Schles.* 3 (1): 688 (1889).
Sphaerobolus dentatus (With.) W. G. Smith, *Brit. Basid.*: 485 (1908).
Sphaerobolus terrestris (Alb. & Schwein.) W. G. Smith, *Brit. Basid.*: 486 (1908).

Selected descriptions: Buller (1933); Cejp (1958: 674, figs. 253–254); Ingold (1972); Ramsbottom (1953: 254, pl.14c); Rea (1922: 54); Wakefield & Dennis (1981: 211, pl.111/5).

Selected illustrations: Berkeley (1860 : 312, pl.2/2); Breitenbach & Kränzlin (1986: pl.498); Gerhardt (1985: 186); Jahn (1979: 232, pl. 208); Massee (1889: 60, fig. 55); Phillips (1982: 255).

Diagnostic characters: minute, clustered, whitish to orange fruitbodies; stellate, apical dehiscence; solitary, dark peridiole.

Fruitbody minute, 1–1.5 mm diam., globose, sessile, partly buried in a white mycelial mat. *Peridium* four-layered; outer layer whitish to orange, smooth, with stellate dehiscence forming 6–8 rays at apex on maturity to expose the peridiole; inner peridial wall orange, finally everting to eject peridiole. *Peridiole* solitary, globose, reddish brown, shiny, embedded in mucilage. *Epiphragm* absent.

Basidiospores 6–10 x 5.5–6.5 µm, subglobose to ovo-ellipsoid, hyaline, thick-walled, smooth. *Peridiopellis* hyphae inflated, with clamp-connexions.

Habitat: on rotting wood and decaying plant debris, sawdust, straw, horse, cow and rabbit dung, often forming large groups.

Distribution and frequency: widespread; common throughout the year. Cosmopolitan.

Fig. 50. *Sphaerobolus stellatus*. A, (Denmark, Ångermand, 4 Sept. 1986, Nitare); B, Hampshire, New Forest, Oct. 1990, Dickson).

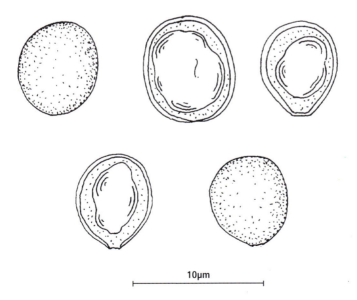

Fig. 51. *Sphaerobolus stellatus* (Lancashire, Formby, 30 Nov. 1963, Palmer 12037). Spores.

PUFFBALLS, BOVISTS AND EARTHSTARS

LYCOPERDALES G.Cunn.
Gasterom. Austr. & New Zeal.: 124 (1944).

Fruitbody sometimes hypogeous in development, epigeous at maturity, globose or subglobose, typically sessile, sometimes with a stem-like base. *Peridium* with two major layers. *Exoperidium* persistent or not. *Endoperidium* thin, dehiscence irregular, or by a pore or pores. *Capillitium* usually present, comprising thin- or thick-walled hyphae, sometimes with pores, branched or not, clamp connexions present or absent. *Gleba* powdery at maturity. *Basidia* cylindric or clavate to capitate, with (1–) 4–8 (–11) sterigmata. *Basidiospores* small, globose to broadly ellipsoid, commonly ornamented with warts or spines, pigmented, pedicellate in some species. *Habitat* terrestrial or rarely lignicolous. *Type family: Geastraceae* Corda.

The order contains the earthstars and puffballs and their allies, and is worldwide in distribution. The largest and most important families, *Geastraceae* and *Lycoperdaceae*, are equally widespread and are well represented in Britain and Europe. Two other families, *Arachniaceae* and *Mesophelliaceae*, are included in the order by Dring (1973). The former includes a single genus, *Arachnion*, with small, hypogeous fruitbodies, fragile peridium and chambered gleba. *Mesophelliaceae* as defined by Dring includes three genera with thick, indehiscent peridium and radially-structured gleba. They are known mostly from the southern hemisphere and are unrecorded from Europe.

Key to British Families

1. Exoperidium well-developed, thick and persistent, 3 (–4)–layered, splitting at maturity, usually in a stellate pattern, to expose the endoperidium. Endoperidium dehiscent by a pore or pores. Capillitium aseptate, typically unbranched
.... **Earthstars** (*Geastraceae*)
1. Exoperidium thin, often inconspicuous, comprising a single layer, commonly lost at maturity. Endoperidium dehiscent by a pore or by irregular breakdown. Capillitium septate or not, typically branched .... **2. Puffballs** (*Lycoperdaceae*)

1. GEASTRACEAE Corda
in *Anleit. Stud. Mykol.*: 104 (1842, as *Geastrideae*).

Fruitbody often hypogeous in development, epigeous at maturity, globose or ellipsoid to umbonate when unexpanded, sessile or rarely stipitate. *Peridium* duplex; exoperidium thick, persistent, 3(4)–layered, splitting stellately at maturity to expose the endoperidial body. *Endoperidium* persistent or not, usually thin but tough, dehiscing by a pore or pores. *Columella* present, rarely multiple. *Capillitium* present, of thin- or thick-walled hyphae, radiating from the columella or rarely enclosed in radiating peridioles. *Basidia* clavate to saccate or cylindric, sterigmata variable in number. *Basidiospores* globose to broadly ellipsoid, up to 10 μm diam., ornamented, guttulate. *Habitat* humicolous, rarely lignicolous or termiticolous. *Type genus*: *Geastrum* Pers.: Pers.

The family is almost cosmopolitan in distribution. Eight genera were recognised by Sunhede (1989), including *Radiigera* Zeller, tentatively referred here following Askew

& Miller (1977) rather than to *Mesophelliaceae* as accepted by Dring (1973). Species of *Geastraceae* are typically distinctive in having stellate splitting of the exoperidium at maturity, but *Pyrenogaster* Malençon & Riousset, as well as *Radiigera* are non-stellate. Genera are separated largely on glebal structure, the form of the capillitium, and characters of the exo- and endoperidium. Spore characters are also of value. Most species of the family are humicolous, in various habitats but commonly on well-drained, calcareous soil (Sunhede 1989). They are saprotrophs, although there are unconfirmed reports of ectomycorrhizal development in a few species.

A detailed study of the morphology, ecology, distribution and taxonomy of *Geastraceae* is given by Sunhede (1989), with full discussion of northern European species.

Key to the British Genera

1. Endoperidial body sessile or borne on a single stalk, unistomatous, containing a single columella .... **1. Earthstars** (*Geastrum*)
1. Endoperidial body borne on several stalks, multistomatous, containing several columellae .... **2. Pepper Pots** (*Myriostoma*)

1. EARTHSTARS

GEASTRUM Pers.: Pers.,
Synops. Meth. Fung.: 131 (1801); *Neues Mag. Bot.* 1: 85 (1794).

Geaster P.Micheli, *Nova Plant. Gen.* p. 220 (1729).
Geaster P.Micheli: Fr., *Syst. Mycol.* 3: 8 (1829).

Fruitbody epigeous or hypogeous in development, sessile (in British species), subglobose to obovate or lageniform, sometimes umbonate, with basal mycelium; exoperidium present, splitting at maturity from the apex and turning downwards as individual, pointed rays to expose the endoperidium and give a stellate appearance, hygroscopic in some species. *Exoperidium* of 3 or 4 layers, comprising an outermost mycelial layer which may bind debris, a central fibrous layer and an innermost pseudoparenchymatous layer. A further layer of loosely woven, thin-walled hyphae including calcium oxalate crystals is present between the pseudoparenchymatous layer and the endoperidium in some species. *Mycelial layer* composed of thin- or thick-walled, interwoven hyphae. *Fibrous layer* comprising usually thick-walled, densely interwoven hyphae. *Pseudoparenchymatous layer* cellular, composed of either thin- or thick-walled cells. *Endoperidial body* sessile or stipitate, attached to the fibrous layer of the exoperidium, with apical peristome and central columella. *Endoperidium* well developed, persistent, smooth or roughened with small warts or ridges. *Peristome* variable in form, usually conical, striate or sulcate to plicate, or simply fibrillose, sometimes distinctly delimited by a smooth area or circular ridge. *Columella* single, central, basally attached to the fibrous layer, variable in form, commonly cylindric or clavate.

Capillitium composed of simple, unbranched, rarely forked, usually thick-walled, smooth or ornamented hyphae which lack clamps and radiate from the columella. *Basidia* clavate to cylindric, sometimes stalked, usually less than 35 x 10 µm, with (1–) 4–8 (–11) sterigmata, with basal clamp-connexion, epibasidium sometimes present.

Basidiospores 3–8 μm diam., globose, thin-walled, usually verruculose, yellow-brown to dark brown. *Habitat* terricolous on soil or amongst litter, especially on calcareous soils on well drained sites (Sunhede 1989), summer to winter; saprophytic, possibly mycorrhizal. *Distribution* cosmopolitan, in tropical and temperate regions excluding Antarctica. *Type species: Geastrum multifidum* Pers. (= *G. coronatum* Pers.: Pers.)

The earlier name *Geaster* Micheli has been commonly, though not universally, considered as an orthographic variant of the sanctioned *Geastrum* Pers.: Pers. Nomenclature and typification of the genus was investigated by Demoulin (1984), who has shown these names to be homotypic and that *Geastrum* is to be treated as an orthographic variant. The type species, *G. multifidum* Pers., was shown by Demoulin to be an obligate synonym of *G. coronatum* Pers.: Pers., both typified by the same figure (Schmidel, 1793 pl. 46). The genus is distinctive with pointed lobes which turn back to create a stellate pattern from which the popular name earthstar is derived. The extent of development of the mycelial layer varies according to whether the species is hypogeal or epigeal in development. Initially, hypogeal species, such as *G. rufescens*, have a well-developed mycelial layer which binds debris.

Key to British species

1. Peristome plicate 2
1. Peristome fimbriate 7

   2. Endoperidial body sessile, 6–11 mm diam .... **1. Elegant Earthstar** (*G. elegans*)
   2. Endoperidial body with distinct stalk when dry, 5–35 mm diam 3

3. Rays hygroscopic; endoperidium verrucose .... **2. Field Earthstar** (*G. campestre*)
3. Rays not hygroscopic; endoperdium verrucose or smooth 4

   4. Endoperidium verrucose .... **3. Berkeley's Earthstar** (*G. berkeleyi*)
   4. Endoperidium smooth or pruinose 5

5. Small species; unexpanded fruitbody subglobose or slightly umbonate; endoperidial body 12 mm or less in diam.; mostly on sand dunes. ....
.... **4. Dwarf Earthstar** (*G. schmidelii*)
5. Larger species; unexpanded fruitbody commonly umbonate; endoperidial body (10) 15–27 mm diam.; in other habitats 6

   6. Endoperidial body with basal collar; endoperidium with fine, innate horizontal striae .... **5. Striated Earthstar** (*G. striatum*)
   6. Endoperidial body without basal collar; endoperidium without innate striae ....
.... **6. Beaked Earthstar** (*G. pectinatu*m)

7. Mycelial layer encrusting debris; fruitbody subglobose, not or only slightly umbonate.
.... 8
7. Mycelial layer not encrusting debris; fruitbody onion-shaped, distinctly umbonate ... 14

   8. Endoperidial body sessile 9
   8. Endoperidial body with distinct stalk when dry 10

9. Hygroscopic; spores dark brown, 6–7.5 µm diam. including ornament of irregular warts; endoperidium scurfy, not puberulent .... **7. Daisy Earthstar** (*G. floriforme*)

9. Not hygroscopic; spores pale, 3.3–4 µm diam. including ornament of fine isolated warts; endoperidium finely puberulent .... **8. Sessile Earthstar** (*G. fimbriatum*)

10. Fruitbody fornicate 11

10. Fruitbody not fornicate 12

11. Spores 3.8–4.5 µm diam. including ornament; endoperidial body 15–25 mm diam.; endoperidium finely puberulent .... **9. Arched Earthstar** (*G. fornicatum*)

11. Spores 5.5–6.3 µm diam. including ornament; endoperidial body 7–13 mm diam.; endoperidium pruinose, not puberulent
.... **10. Four-rayed Earthstar** (*G. quadrifidum*)

12. Small; endoperidial body 7–10 mm diam., expanded fruitbody 1.5–3 cm across spores 6–7.5 µm diam. including ornament .... **11. Tiny Earthstar** (*G. minimum*)

12. Larger; endoperidial body 11–34 mm diam., expanded fruitbody 3–10 cm across; spores 4.2–6.5 µm diam. including ornament 13

13. Pseudoparenchymatous layer becoming pinkish; endoperidium finely puberulent with erect, thick-walled hairs; spores 4.2–5 µm diam. including ornament
.... **12. Rosy Earthstar** (*G. rufescens*)

13. Pseudoparenchymatous layer not becoming pinkish; endoperidium smooth or felty but without erect hairs; spores 5–6.5 µm diam. including ornament
.... **13. Crowned Earthstar** (*G. coronatum*)

14. Fruitbody strongly hygroscopic; mycelial layer not splitting radially.
.... **14. Weathered Earthstar** (*G. corollinum*)

14. Fruitbody not hygroscopic; mycelial layer splitting radially or not 15

15. Endoperidial body 21–39 mm diam. endoperidium glabrous; pseudoparenchymatous layer often splitting to form a collar around the endoperidial body; mycelial layer usually without conspicuous radial splits .... **15. Collared Earthstar** (*G. triplex*)

15. Endoperidial body 9–19 mm diam.; endoperidium, at least when young, finely puberulent with erect, thick-walled hairs; pseudoparenchymatous layer not splitting to form a collar; mycelial layer commonly with conspicuous radial splits
.... **16. Flask-shaped Earthstar** (*G. lageniforme*)

1. ELEGANT EARTHSTAR

Map No. 24

Geastrum elegans Vittad., *Monogr. Lycoperd.*: 15 (1842, as *Geaster*).

Selected descriptions: Sunhede (1989: 164–179).

Selected illustrations: Sunhede (1989: figs 63–66).

Diagnostic characters: peristome fimbriate; rays frequently recurved below; spore-sac small, sessile; spores 4.8–5.7 μm (excl. orn.).

Fruitbody hypogeous, subglobose or ovate, encrusted with debris. Expanded fruitbody 12–35 mm across, splitting to about halfway into 6–8 rays, non-hygroscopic, rays arched or spreading, sometimes recurved under the fruitbody. *Mycelial layer* persistent, sometimes peeling off, strongly encrusted with soil and debris. *Fibrous layer* whitish to yellowish. *Pseudoparenchymatous layer* at first whitish to pale brown, becoming dark brown, usually splitting and peeling with age, sometimes forming a collar around the endoperidial body. *Endoperidial body* sessile, subglobose to ovate, without an apophysis, 6–11 mm diam. *Endoperidium* pale brown or greyish, glabrous, often with a whitish, crystalline pruina. *Peristome* conical, 1.5–3 mm high, plicate with 14–20 distinct ridges, clearly delimited but usually not bounded by a circular ridge. *Columella* whitish, narrowly clavate. *Mature gleba* dark brown.
  Basidia not seen. *Capillitial hyphae* 3–7 μm diam., gradually tapered to rounded tips 1.5–4 μm diam., yellowish brown, thick-walled, with a narrow, discontinuous lumen, surface irregularly encrusted. *Spores* globose, yellow-brown to brown, verruculose, 4.8–5.7 μm diam. excluding ornament, 5.6–7 μm diam. including ornament, verruculae 0.4–0.6 μm high, often coalescent, 0.3–2 μm wide.

Habitat: on sandy soil, usually coastal. It occurs in other habitats, on calcareous soil, in Europe, and its ecology is discussed in detail by Sunhede (1989).

Distribution and frequency: Evidently very rare in the British Isles, and perhaps now extinct here. Only a single confirmed collection has been made this century and only four collections have been available for study, as follows: Isles of Scilly, St. Martins, The Plains, 4 Dec. 1947, Gregory 777 (K); Norfolk, Great Missingham, 6 Oct. 1892, ex Herb. Plowright (K); Unlocalized, undated, on 'sand near the sea', Fergusson, ex Herb. Broome (K); Unlocalized, undated, ex Herb. Plowright (K). It is widely distributed in Europe, but with a southern distribution according to Sunhede (1989)

Other remarks: Fruitbodies of *Geastrum elegans* may be similar in general appearance to those of the much more common *G. fimbriatum*, but they can be readily distinguished by their plicate peristome and larger, more coarsely ornamented spores. The species requires more careful distinction from *G. schmidelii* which may occur in similar habitats and cannot be distinguished on spore characters. *Geastrum elegans* differs particularly in having the endoperidial body sessile and without an apophysis. In addition, the rays are often recurved beneath the fruitbody in *G. elegans*, and the peristome of this species is commonly slightly darker than the surrounding endoperidium and is not bounded by a circular ridge.

Elegant Earthstar

Fig. 52. *Geastrum elegans*. (Sweden, Gotland, Bunge Parish, Sunhede).

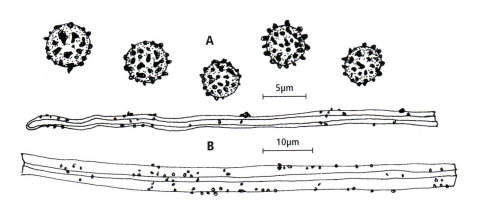

Fig. 53. *Geastrum elegans*. (No locality, ex-Herb. Broome). A, Spores; B, capillitial threads.

2. FIELD EARTHSTAR

Map No. 25

Geastrum campestre Morgan in *Amer. Naturalist* 21: 1027 (1887, as *Geaster*).
Geastrum pseudomammosum Henn. in *Hedwigia* 39, *Beibl.*: (54) (1900, as *Geaster*).
Geastrum asperum Lloyd, *Mycol. Not.* 7: 70 (1901, as *Geaster*).

Selected descriptions: Sunhede (1989: 111–129).

Selected illustrations: Sunhede (1989: figs 37–42).

Diagnostic characters: peristome plicate; spore-sac with a distinct stalk; rays hygroscopic; endoperidium verrucose.

Fruitbody hypogeous in development, subglobose to depressed, not umbonate, encrusted with debris. Expanded fruitbody c. 20–50 mm across, exoperidium splitting to about half way into 7–10 rays, hygroscopic, rays spreading, usually arched when fresh and recurved at the tips, bending upwards onto the endoperidial body when dry. *Mycelial layer* persistent, binding debris. *Fibrous layer* whitish to pale grey-brown, not often exposed. *Pseudoparenchymatous layer* fleshy, at first pale brown, becoming darker with age, eventually dark brown and often longitudinally wrinkled, persistent, usually not cracking or peeling. *Endoperidial body* subglobose or obovate, 12–14 mm across, short-stipitate. *Stalk* c. 1–2 mm high, whitish to brown, often flattened in cross-section. *Endoperidium* pale grey-brown, usually densely covered in small warts which are often more prominent around the peristome, with age becoming worn and almost smooth. *Peristome* conical, strongly plicate with 15–20 grooves, clearly delimited, usually with a slight, circular ridge. *Columella* hemispherical to cylindric, pale brown. *Mature gleba* dark brown.

Basidia not seen. *Capillitial hyphae* pale yellow-brown to almost hyaline, finely tapered, occasionally forked or lobed, thick-walled with a narrow lumen, smooth or partially encrusted. *Spores* globose, yellow-brown to brown, verruculose, (4.5–) 4.8–5.7 µm diam. excluding ornament (British collections), 5.6–6.7 µm diam. including ornament, verruculae isolated, irregular in outline, 0.3–0.6 µm high, 0.3–1.2 µm across.

Habitat: in parks and gardens; on calcareous, well-drained soil (Sunhede 1989).

Distribution and frequency: only two collections are known from the British Isles, viz.: Surrey, Kew, Royal Botanic Gardens, Director's garden, 29 July 1926 (K); Kent, Ickleford churchyard, 3 Dec. 1958, Edwards (K). The species is probably not native to Britain, but occasionally introduced. It occurs throughout Europe, but is very rare in northern parts (Sunhede 1989), and is clearly widely distributed in north temperate regions, being described from the U.S.A.

Other remarks: This species seems to vary noticeably with regard to size and colour of the mature spores, and coarseness of the warts. The spore characters in each of the two British collections are similar, although the spores of the Kew collection are somewhat darker and bear slightly coarser warts. Other collections examined here have somewhat larger spores, although apparently none as large as 8 µm diam. given by Sunhede (1989). In isotype material (U.S.A., Nebraska, roadsides at Lincoln, 27 Sept. 1886, Bessey, Ellis & Everhart, *North American Fungi* ser. 2, 1940) the spores are rather pale yellow-brown and measure 5.2–6.5 µm diam. excluding ornament, 6–7.3 µm including ornament, which varies from fine to rather coarse warts. However, other characters in each of these collections are comparable and it seems likely that only a single species is involved. *Geastrum ambiguum* Morgan is probably also conspecific. *Geastrum campestre* is apparently closely related to *G. berkeleyi* and differs only in having smaller, hygroscopic fruitbodies.

Fig. 54. *Geastrum campestre.* (Denmark, Skåne, 13 May 1990, Nitare).

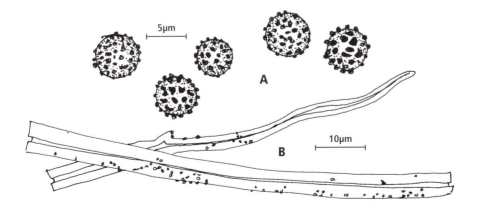

Fig. 55. *Geastrum campestre.* (Kent, Icklefield, Dec. 1958, Edwards). A, Spores; B, capillitial threads.

3. BERKELEY'S EARTHSTAR

Map No. 26

Geastrum berkeleyi Massee in *Ann. Bot. (London)* 4: 79 (1889, as *Geaster*).
Geastrum pseudostriatum Hollós in *Math. Természettud. Értes.* 19: 505 (1901, as *Geaster*).
Geastrum hollosii V.J.Stanek in Pilát, *Flora CSR, B.1 Gasteromycetes*: 467 (1958).

Selected descriptions: Sunhede (1989: 88–111).

Selected illustrations: Dörfelt (1985: figs 6/3, 30, 31); Ellis (1981: fig. 1/11); Rea (1912: pl. 18, as *G. asper*); Ryman & Holmåsen (1984: 600); Sunhede (1989: figs 23–34).

Diagnostic characters: relatively large fruitbody; non-hygroscopic exoperidium; plicate peristome, with smooth surrounding area; stalked spore-sac; verruculose endoperidium.

Fruitbody with hypogeous development, at first subglobose, 1–5 cm across, encrusted with debris. Expanded fruitbody 4–8 (–10) cm across; exoperidium splitting into 5–9 rays which become reflexed and usually arched downwards, not hygroscopic although sometimes partly curled back when dry. *Mycelial layer* well developed, binding debris and soil particles. *Pseudoparenchymatous layer* whitish at first, becoming brown to dark brown or reddish brown, tending to fragment and peel off in patches. *Endoperidial body* stipitate, subglobose to pyriform, 13–35 mm diam. in the type, often constricted below and with an apophysis. Stipe short, 1–3 mm high (dried material), pale brownish. *Endoperidium* brown to grey-brown, finely but conspicuously roughened with minute, irregular verruculae < 1 mm across except for a smooth, circular area surrounding the peristome. *Peristome* conical, 3–5.5 mm high, distinctly plicate with c. 25–30 folds. *Columella* depressed to clavate or subglobose. *Mature gleba* dark purplish brown.

 Basidia not seen. *Spores* purple-brown in mass, globose, distinctly verruculose, 4.7–6 μm diam. excluding ornament, 5.5–7 μm diam. incl. ornament, verruculae isolated, 0.3–0.5 μm high, 0.3–0.8 μm across. *Capillitial hyphae* yellowish brown, thick-walled, with lumen often partially obscured, mostly 5–9 μm diam., tapered gradually to subacute tips, surface smooth or encrusted.

Habitat: on well-drained, calcareous soil, in coniferous or deciduous woodlands or in open sites (Sunhede 1989).

Distribution and frequency: very rare and possibly extinct in Britain. Known with certainty only from a few collections, notably from Nottinghamshire and Norfolk. Elsewhere, recorded from Denmark and Sweden.

Other remarks: Ellis (1981) reported material from Norfolk collected in 1925 or 1926, not seen in the present study. Palmer (1968) cited an unconfirmed Scottish record, but few other confirmed British collections more recent than the type (Nottinghamshire, Lambley, ex Herb. Hookerianum, lectotype & isolectotype, K) are known.

 The species was redescribed and discussed in detail by Sunhede (1989), who designated the lectotype. The lectotype collection is in good condition, and shows the characters of the species clearly. Massee (1889), in the protologue, cited 'Ascot; Lambley, Notts.; Laxton. Type - Herb. Berk. 4550'. Only the Lambley collection is preserved in K, although there are 4 other packets of unlabelled material. However, the cited type and other collections may be available in the Massee herbarium which is preserved in New York (NY).

Berkeley's Earthstar

Fig. 56. *Geastrum berkeleyi.* (Sweden, Gotland, Rute Parish, Sunhede).

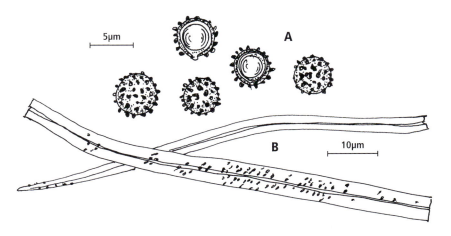

Fig. 57. *Geastrum berkeleyi.* (Nottinghamshire, Lambly, Herb. Berk., lectoype K). A, Spores; B, capillitial threads.

British Puffballs, Earthstars and Stinkhorns

4. DWARF EARTHSTAR

Map No. 27

Geastrum schmidelii Vittad., *Monogr. Lycoperd.*: 13 (1842, as *Geaster*).
Geastrum nanum Pers., *J. Bot. (Morot)* 2: 27 (1809, nom. illegit.).

Selected descriptions: Sunhede (1989: 397–415).

Selected illustrations: Brand (1988: 111); Dörfelt (1985: figs 44, 45, as *G. nanum*); Ellis (1981: figs 1/12 & 3/7, as *G. nanum*); Michael, Hennig & Kreisel (1986: fig 162); Phillips (1981: 253, as *G. nanum*); Sunhede (1989: figs 177–185).

Diagnostic characters: fruitbody small, subglobose; on sand dunes; peristome plicate; rays not hygroscopic; spore-sac less than 12 mm, smooth, stalked.

Fruitbody hypogeous at first, subglobose to depressed or ovate, rarely umbonate, strongly encrusted with sand and debris. Expanded fruitbody 14–37 mm across, rarely larger, splitting to about half way into 5–8 rays, rays non-hygroscopic, arched and somewhat recurved or almost horizontal. *Mycelial layer* present, persistent, strongly encrusting sand and debris. *Fibrous layer* papery, whitish to pale yellowish brown. *Pseudoparenchymatous layer* whitish at first, becoming pale brown, darkening with age and usually dark brown in dried specimens. *Endoperidial body* short-stalked, subglobose to ovate or depressed, 5–12 mm diam., with an apophysis; stalk 1–2 mm high (dried), whitish. *Endoperidium* grey-brown to clay-coloured, often minutely whitish-pruinose when fresh, becoming smooth. *Peristome* usually distinctly delimited by a low, circular ridge and paler than the surrounding endoperidium, usually acutely conical, 2–4 mm high, plicate, mostly with 14–20 ridges. *Columella* small and rather weakly developed, rounded to narrowly clavate. *Mature gleba* dark brown.
 Basidia not seen. *Capillitial hyphae* 2–6 µm diam., pale brown to yellow-brown or almost hyaline, thick-walled with a narrow, discontinuous lumen, gradually and finely tapered to rounded or subacute tips 1–2.5 µm diam., irregularly encrusted. *Spores* dark brown, globose, verruculose, 4.8–5.6 µm diam. excluding ornament, 5.6–6.5 µm diam. including ornament, verruculae irregular, often coalescent, 0.3–0.7 µm high, 0.3–1.5 µm across.

Habitat: on coastal sand dunes, amongst grass and moss in dune slacks, occasionally elsewhere in calcareous, sandy places. Other habitats in Europe discussed in detail by Sunhede (1989).

Distribution and frequency: Widespread and fairly common in appropriate habitats throughout England and Wales, but scarce in Scotland. North temperate in distribution.

Other remarks: this is a small, distinctive earthstar, recognized by its plicate, well-delimited peristome, apophysate, short-stalked endoperidial body, persistent mycelial layer which strongly encrusts sand and debris, and habitat. *Geastrum pectinatum* is similar in many respects but is larger, with a longer, more distinct stalk to the endoperidial body, and an endoperidium which, at least when fresh, is covered with a thick crystalline pruina. It also differs in habitat. *Geastrum elegans* is also similar to *G. schmidelii*, and, though a much rarer species, requires careful comparison. The differences between these species are discussed under *G. elegans*.

Dwarf Earthstar

Fig. 58. *Geastrum schmidelii*. (Denmark, Vandplasken, 18 Oct. 1988, Laessøe).

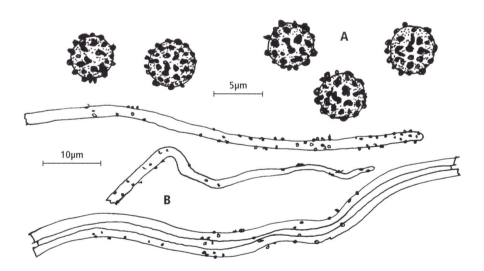

Fig. 59. *Geastrum schmidelii*. (Cheshire, Wirrel, Oct. 1952, Palmer 175). A, Spores; B, capillitial threads.

5. STRIATED EARTHSTAR

Map No. 28

Geastrum striatum DC., *Fl. franç.* ed. 3, 2: 267 (1805).
Geastrum bryantii Berk. in Smith, *Engl. fl.* 5: 300–301 (1836, as *Geaster*).
Geastrum orientale Hazsl in *Grevillea* 6: 108–109 (1878, as *Geaster*).

Selected descriptions: Sunhede (1989: 426–445).

Selected illustrations: Dörfelt (1985: figs 5/5, 6/4, 60, 61); Ellis (1981: figs 1/8–9 & 3/12); Michael, Hennig & Kreisel (1986: fig 161); Sunhede (1989: figs 192–199).

Diagnostic characters: peristome plicate; rays not hygroscopic; spore-sac with basal collar, stalked, surface finely striate.

Fruitbody hypogeous, subglobose, often umbonate or onion-shaped, strongly encrusted with soil and debris. Expanded fruitbody 28–65 mm across, exoperidium splitting to about half way into 6–9 rays, non-hygroscopic, rays arched, somewhat recurved or spreading horizontally. *Mycelial layer* persistent, strongly encrusting soil and debris. *Fibrous layer* whitish to dull grey-brown. *Pseudoparenchymatous layer* fleshy, sheathing the stalk when fresh, whitish at first, becoming fawn to cigar-brown, often splitting, soon peeling away from the stalk. *Endoperidial body* 10–20 (–26) mm diam., prominently stalked, subglobose or depressed, usually with an apophysis and with a distinct basal collar. *Stalk* 3–6 mm high, tapered downwards, pale grey-brown to brown, often sheathed at the base by a ring-like zone of pseudoparenchymatous tissue. *Endoperidium* covered with a felty or mealy, whitish to pale fawn layer which is gradually lost with age, surface beneath this layer often reddish brown and marked with fine striae. Striae horizontal, becoming vertical towards the base of the endoperidium, especially distinct when rubbed. *Peristome* clearly delimited, conical, 2–4 mm high, plicate, with mostly 20–25 ridges. *Columella* whitish or pale brown, rounded to conical. *Mature gleba* dark brown, with a purplish tinge.

Basidia not seen. *Capillitial hyphae* 3–8 µm diam., pale yellow-brown, tapered gradually to rounded tips, occasionally forked, thick-walled, with narrow lumen, smooth or partially finely encrusted. *Spores* globose, dark yellow-brown, verruculose, 3.8–4.8 µm diam. excluding ornament, 5–6 µm diam. including ornament, verruculae fairly coarse, irregular and sometimes coalescent, 0.3–0.7 µm high, 0.4–1.5 µm across.

Habitat: in woods, gardens and parks, amongst litter, often on rich or calcareous soil; usually gregarious.

Distribution and frequency: widely distributed and fairly frequent in the British Isles, and throughout Europe.

Other remarks: the interpretation of this species, for which no type material is known to exist, is discussed by Palmer (1968) and Sunhede (1989). The basal endoperidial collar is not mentioned in the original description by De Candolle (1805), who refers, however, to an illustration by Bryant (1782) in which the collar is clearly shown. Examination of cotype material in K of *G. bryantii* (Suffolk, Bungay, leg. Stock) confirms its identity with this concept of *G. striatum* suggested by various authors including Palmer and Sunhede.

The basal collar to the endoperidial body is a diagnostic character of *G. striatum*, and the most obvious character distinguishing it from the very similar *G. pectinatum*. A ring-like zone is often also present at the base of the stalk. There is a felty covering to the endoperidium in fresh material, which is characteristic though similar to that of *G. pectinatum*. This is gradually abraded with age, and beneath it the endoperidial surface is finely but distinctly streaky or striate, a character which may also be diagnostic of the species.

This species is evidently closely related to *G. pectinatum*, and is difficult to separate on microscopic characters. However, the capillitial hyphae are generally somewhat paler than those of *G. pectinatum*, and the spores, though similar in size, have less elevated warts.

Fig. 60. *Geastrum striatum*. (Denmark, Samsø. 1991, Lange).

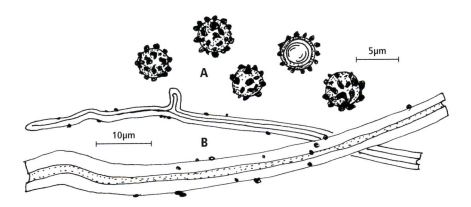

Fig. 61. *Geastrum striatum* . (Suffolk, Bungay, type of *G. bryantii*). A, Spores; B, capillitial threads.

6. BEAKED EARTHSTAR

Map No. 29

Geastrum pectinatum Pers., *Synops. Meth. Fung.*: 132 (1801).
Geastrum plicatum Berk. in *Ann. Mag. Nat. Hist.* 3: 399 (1839, as *Geaster*).
Geastrum tenuipes Berk. in Hook., *London J. Bot.* 7: 576 (1848, as *Geaster*).

Selected descriptions: Breitenbach & Kränzlin (1986: 499); Marchand (1976: p. 126); Sunhede (1989: 294–312).

Selected illustrations: Breitenbach & Kränzlin (1986: 499); Cetto (1988: 364); Dörfelt (1985: figs 5/6, 46, 47); Ellis (1981: figs 1/10 & 3/8); Marchand (1976: pl. 361); Michael, Hennig & Kreisel (1986: fig 160); Ryman & Holmåsen (1984: 600); Sunhede (1989: figs 124–130).

Diagnostic characters: peristome plicate; unexpanded fruitbody umbonate; rays not hygroscopic; spore-sac smooth, lacking striae; no collar.

Fruitbody at first hypogeous, subglobose, commonly umbonate, encrusted with debris. Expanded fruitbody 29–75 mm across, exoperidium splitting to about half way into 6–9 rays, non-hygroscopic, arched, rays spreading horizontally or somewhat recurved. *Mycelial layer* present, usually persistent, strongly encrusting soil and debris. *Fibrous layer* whitish or pale yellowish-brown, denuded with age. *Pseudoparenchymatous layer* fleshy when fresh, sheathing the stalk, whitish at first, becoming brown, dark brown when old, often splitting and partly peeling, usually shrinking around the stalk when dry to leave a thickened, ring-like zone. *Endoperidial body* distinctly stalked, 15–27 mm diam., subglobose or depressed, with or without an apophysis, frequently ridged or striate on the underside and tapered towards the stalk. *Stalk* cylindric or compressed, 3–5 mm high, whitish or grey-brown when denuded. *Endoperidium* with a pale brown, mealy covering when fresh, when abraded becoming grey-brown to dark brown or lead-grey, with a whitish pruina, otherwise smooth. *Peristome* conical, 2–5 mm high, usually strongly plicate, with 20–32 ridges, clearly delimited but not bounded by a circular ridge. *Columella* usually narrowly conical and protruding more than half way into the gleba, whitish or pale brown. *Mature gleba* dark brown.
   Basidia not seen. *Capillitial hyphae* 3–7 μm diam., dark yellow-brown to brown, tapered, simple or occasionally forked near the tips, thick-walled with a narrow lumen, smooth or slightly encrusted. *Spores* globose, dark yellow-brown, coarsely verruculose, 4.2–4.8 μm diam. excluding ornament, 6–7 μm diam. including ornament, verruculae cylindric or clavate, blunt, 0.6–1.2 μm high, sometimes coalescent, 0.4–1.5 μm across.

Habitat: usually associated with conifers, in woods or parks or at roadsides, occasionally with deciduous trees; often gregarious, fruiting in autumn.

Distribution and frequency: widely distributed and fairly frequent in England and Wales, but apparently unrecorded from Scotland. One of the commonest earthstars in northern Europe according to Sunhede (1989).

Other remarks: The coarsely ornamented spores are fairly distinctive. *Geastrum schmidelii* may be similar in appearance, but is smaller, has a shorter stalk, slightly larger but less coarsely ornamented spores, and occurs in different habitats. *Geastrum striatum* is also similar to *G. pectinatum* but differs most obviously in the basal collar to the endoperidial body.

Beaked Earthstar

Fig. 62. *Geastrum pectinatum*. (Germany, Sanhausen, 5 Nov. 1972, Winterhoff).

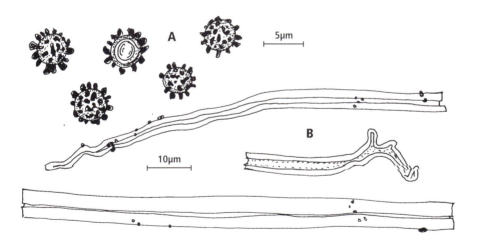

Fig. 63. *Geastrum pectinatum*. (Surrey, Fetcham, Parker). A, Spores; B, capillitial threads.

7. DAISY EARTHSTAR

Map No. 30

Geastrum floriforme Vittad., *Monogr. Lycoperd.*: 23 (1842, as *Geaster*).
Geastrum delicatum Morgan in *Amer. Natural.* 21: 1028 (1887, as *Geaster*) *teste* Stanek (1958).
Geastrum pazschkeanum Henn. in *Hedwigia* 29: 55 (1900, as *Geaster*) *teste* Stanek (1958).
Geastrum sibiricum Pilát in *Bull. Soc. Mycol. Fr.* 51: 423 (1935) *teste* Stanek (1958).

Selected descriptions: Sunhede (1989: 198–209).

Selected illustrations: Dörfelt (1985: fig 36); Ellis (1981: figs 1/6–7 & 3/5); Michael, Hennig & Kreisel (1986 fig 174); Sunhede (1989: figs 77–81).

Diagnostic characters: peristome fimbriate; fruitbody hygroscopic; mycelial layer encrusting debris; endoperidium sessile, scurfy; spores 6–7.5 µm, dark brown.

Fruitbody hypogeous in development, subglobose, 10–22 mm across, encrusted with soil and debris. Expanded fruitbody 18–45 mm across, exoperidium splitting into 6–10 rays, strongly hygroscopic, somewhat recurved under the fruitbody when damp, turning back to cover the endoperidial body when dry. *Mycelial layer* thin, encrusting debris but readily peeling off, leaving whitish remnants. *Fibrous layer* greyish brown, exposed when the mycelial layer is lost. *Pseudoparenchymatous layer* 0.5–0.8 mm thick, persistent, whitish or pale brown at first, becoming darker brown with age, hard when dry. *Endoperidial body* sessile, globose or subglobose, 5–15 mm diam. *Endoperidium* pale greyish brown, minutely scurfy or furfuraceous. *Peristome* fibrillose, not distinctly delimited. *Columella* whitish, cylindric, not clearly visible at maturity. *Mature gleba* dark brown.

Basidia not seen. *Capillitial hyphae* 4–7 µm diam., pale yellow-brown to brown, thick-walled, with a narrow lumen, smooth or irregularly encrusted, gradually tapered to subacute tips. *Spores* globose, (5–) 5.5–7 µm diam. excluding ornament, 6–7.5 µm including ornament, dark brown, with a single guttule, verruculose, verruculae irregular and often coalescent, 0.2–0.5 µm high.

Habitat: usually on well-drained soil. British collections have been associated with *Cupressus* and *Quercus ilex*, but no special association with these trees was noted by Sunhede (1989), who provides a full discussion of the known ecology of this species.

Distribution and frequency: this is a rare species in Britain, having been first collected in Lancashire in 1952 (Palmer 1952, 1955). However, it appears to have become more frequent in recent years, with several collections having made since 1985. It tends to occur more often in coastal localities, but is also found inland. Now also known from Cornwall, Essex, East Sussex, Norfolk and Surrey. Occurs throughout northern Europe.

Other remarks: fruitbodies of *G. floriforme* are gregarious, and may occur in large numbers; up to 40 were noted in a collection from Seaford, East Sussex. *Geastrum floriforme* is closely related to *G. corollinum*, the differences between them being noted above under that species.

Daisy Earthstar

Fig. 64. *Geastrum floriforme*. (Surrey, Kew, Nov. 1990, Brown).

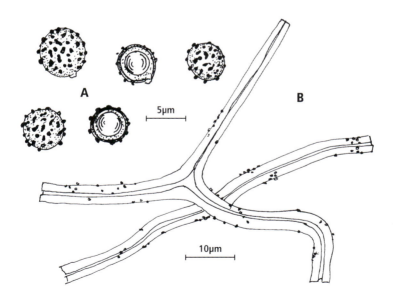

Fig. 65. *Geastrum floriforme*. (Sussex, Seaford, Apr. 1985, Cavadini). A, Spores; B, capillitial threads.

British Puffballs, Earthstars and Stinkhorns

8. SESSILE EARTHSTAR

Map No. 31

Geastrum fimbriatum Fr., *Syst. Mycol.* 3: 16 (1829, as *Geaster*).
Geastrum rufescens Pers.: Pers. var. *minor* Pers., *Syn. Meth. Fung.*: 134 (1801).
Geastrum rufescens emend Kits van Wav. in *Meded. Ned. Mycol. Ver.* 15: 118–121 (1926, as *Geaster*).
Lycoperdon sessile Sow., *Col. fig. Engl. fung., suppl.* : pl. 401 (1809).
Geastrum sessile (Sow.) Pouzar in *Folia Geobot. Phytotax.* 6: 95 (1971).
Geastrum tunicatum Vittad., *Monogr. Lycoperd.*: 18 (1842, as *Geaster*).

Selected descriptions: Breitenbach & Kränzlin (1986: 501); Marchand (1976: p. 128); Sunhede (1989: 180–198).

Selected illustrations: Breitenbach & Kränzlin (1986: 501); Cetto (1988: 362); Dörfelt (1985: figs 6/8, 35, 65); Ellis (1981: figs 2/9–10 & 3/11, as *G. rufescens*); Gerhardt (1985: 207, as *G. sessile*); Marchand (1976: pl. 362); Michael, Hennig & Kreisel (1986 fig 173); Phillips (1981: 252, as *G. sessile*); Sunhede (1989: figs 68–75).

Diagnostic characters: peristome fimbriate; fruitbody not hygroscopic; mycelial layer encrusting debris; endoperidium sessile, puberulent; spores small, 3.3–4 μm, finely verruculose.

Fruitbody hypogeous, subglobose to ovate, sometimes slightly umbonate, encrusted with debris. Expanded fruitbody 21–58 mm across, splitting to about halfway into 5–9 rays, non-hygroscopic, rays arched and commonly recurved under the fruitbody giving a saccate appearance. *Mycelial layer* present, persistent or peeling off, encrusting soil and debris. *Fibrous layer* exposed with age or by separation of mycelial layer when dry, rather papery, whitish to cream. *Pseudoparenchymatous layer* whitish at first, becoming pale brown or yellowish brown, darker with age, often splitting or partly peeling off. *Endoperidial body* 9–25 mm across, subglobose or depressed, without an apophysis, sessile or with a very short stalk. *Endoperidium* pale grey-brown or buff, minutely puberulent with protruding, hyaline or pale yellowish, cylindric-clavate hyphae 30–160 x 10–20 μm, with thickened, refractive walls. *Peristome* fibrillose, not distinctly delimited. *Columella* narrowly clavate, whitish. *Mature gleba* pale brown to brown.
    Basidia not seen. *Capillitial hyphae* 2.5–7 μm diam., finely tapered to the tips, pale yellowish to almost hyaline, walls thickened, smooth or finely encrusted, especially towards the tips. *Spores* globose, yellow-brown, finely verruculose, 3–3.7 μm diam. excluding ornament, 3.3–4 μm diam. including ornament, verruculae fine, isolated, 0.1–0.3 μm high, c. 0.2–0.5 μm across.

Habitat: in deciduous or coniferous woodland, or in parks or gardens associated with trees, usually on calcareous soil.

Frequency and distribution: Widely distributed in the British Isles and comparatively frequent in England and Wales. Apparently rarer in Scotland, but nevertheless one of the commonest earthstars, both in Britain and throughout Europe.

Other remarks: *Geastrum fimbriatum* is distinguished by its sessile or very short-stalked endoperidial body, fibrillose peristome, comparatively pale gleba with pale capillitial hyphae, and small, finely verruculose spores. Differences from similar species with fibrillose peristome, such as *G. rufescens*, are discussed under those species. A full description of fresh British material, with a detailed discussion of the species is given by Palmer (1955).

Fig. 66. *Geastrum fimbriatum*. (Surrey, Fetcham, Laessøe).

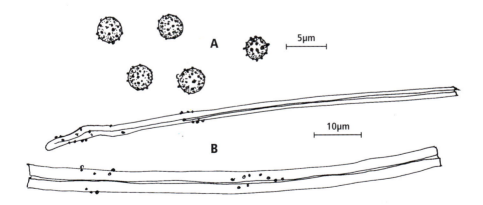

Fig. 67. *Geastrum fimbriatum* . (Lancashire, Freshfield, Dec. 1952, Palmer 87). A, Spores; B, capillitial threads.

9. ARCHED EARTHSTAR

Map No. 32

Geastrum fornicatum (Huds.) Hook., *Fl. Londinensis* 4: 575 (1821).
Lycoperdon fornicatum Huds., *Fl. Angl.*: 502 (1762).
Lycoperdon fenestratum Batsch, *Elench. Fung.*: 151 (1783).
Geastrum fenestratum (Batsch) Lloyd, *Myc. Writings* 1: 70 (1901 as *Geaster*).
Geastrum marchicum Henn. in *Verh. Bot. Vereins Prov. Brandenb.* 34: 4 (1893, as *Geaster*).

Selected descriptions: Marchand (1976: p. 122); Sunhede (1989: 209–225).

Selected illustrations: Dörfelt (1985: figs 6/7, 7/2, 24, 38, 67); Ellis (1981: figs 2/6 & 3/5); Marchand (1976: pl. 359); Michael, Hennig & Kreisel (1986: fig 171); Phillips (1981: 254); Sunhede (1989: figs 82–89).

Diagnostic characters: peristome fimbriate; straight, vertical rays; mycelial layer encrusting debris; spore-sac 15–25 mm, sessile, finely pubescent; spores 3.8–4.5 μm.

Fruitbody hypogeous or partially so in development, globose or depressed, encrusted with debris. Expanded fruitbody 40–80 mm across, distinctly taller than broad, 50–100 mm high, splitting deeply into (3–) 4 (–5) rays, non-hygroscopic, fornicate, rays often almost straight and vertical, with incurved margins and appearing narrow, attached at their tips to the mycelial cup. *Mycelial layer* strongly encrusting soil and debris, cup-like, inner surface whitish to pale brown or grey-brown. *Fibrous layer* pale brown or grey-brown. *Pseudoparenchymatous layer* 3–5 mm thick when fresh, covering the stalk, whitish to pale brown, becoming greyish brown to dark brown with age and when dry, often splitting, hard when dry. *Endoperidial body* subglobose, usually with a distinct apophysis, mostly 15–25 mm across, stalked, stalk 2–3 mm high when dried, pale grey-brown to brown. *Endoperidium* grey-brown, darker with age, sometimes paler at the base, finely puberulent with protruding thick-walled, obtuse or tapered, irregular, pale yellowish hyphae 40–150 x 10–20 μm, and bearing a whitish, crystalline pruina. *Peristome* fibrillose, indistinctly delimited. *Columella* cylindric to conical or clavate, whitish to pale brown. *Mature gleba* dark brown.
    Basidia not seen. *Capillitial hyphae* 6–10 μm diam., thick-walled with narrow, discontinuous lumen, yellow-brown to brown, paler at the tips, gradually tapered, surface smooth or finely encrusted. *Spores* globose, dark yellow-brown to dark brown, finely verruculose, 3.5–4.2 μm diam. excluding ornament, 3.8–4.5 μm diam. including ornament, verruculae fairly regular, 0.2–0.3 μm across, 0.1–0.2 μm high.

Habitat: associated with deciduous trees on rich, well-drained soil.

Distribution and frequency: scarce in Britain. Recorded mainly from southern, central and eastern England, with single records from Wales and Ireland. It is unknown from Scotland. Elsewhere in Europe it is widespread but uncommon or scarce.

Other remarks: this is a distinctive species recognised by its large, fornicate fruitbodies. *Geastrum quadrifidum* is also fornicate but is a consistently smaller species which has larger, more coarsely and irregularly ornamented spores. It appears to be even scarcer in Britain than *G. fornicatum*.

Arched Earthstar

Fig. 68. *Geastrum fornicatum*. (Devon, 1993, Laessøe).

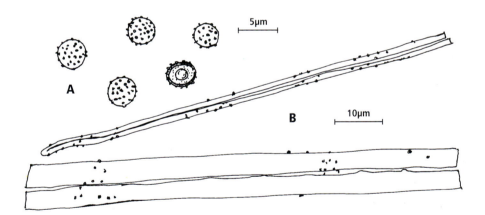

Fig. 69. *Geastrum fornicatum*. (Somerset, Minehead, Oct. 1988, Edwards). A, Spores; B, capillitial threads.

10. FOUR-RAYED EARTHSTAR

Map No. 33

Geastrum quadrifidum Pers.: Pers., *Synops. Meth. Fung.*: 133 (1801); *Neues Mag. Bot.* 1: 86 (1794).
Lycoperdon coronatum Scop., *Fl. carniol.* ed. 2, II: 490 (1772).

Selected descriptions: Breitenbach & Kränzlin (1986: 500); Ellis (1981: 149); Sunhede (1989: 338–356).

Selected illustrations: Breitenbach & Kränzlin (1986: 500); Dörfelt (1985: fig. 53); Ellis (1976: fig. 7); Ellis (1981: figs 2/5 & 3/9); Michael, Hennig & Kreisel (1986: fig 168); Phillips (1981: 253); Ryman & Holmåsen (1984: p.601); Sunhede (1989: figs 146–152).

Diagnostic characters: peristome fimbriate; 4–5 broad, arched rays; mycelial layer encrusting debris; spore-sac 7–13 mm, pruinose, stalked; spores 5.5–6.3 µm.

Fruitbody hypogeous in development, subglobose, encrusted with debris. Expanded fruitbody 13–37 mm across, slightly taller than broad, splitting into 4–5 rays, non-hygroscopic, fornicate; rays broad, arched, with tips attached to mycelial layer. *Mycelial layer* strongly encrusting debris, cup-like, whitish on the inner surface. *Fibrous layer* thin, outer surface whitish to cream or greyish with age. *Pseudoparenchymatous layer* whitish at first, becoming brownish or grey-brown, sometimes with a pinkish tinge, often splitting with age. *Endoperidial body* 7–13 mm across, subglobose to ovate, often taller than broad, stalked (rarely with a double stalk), with an apophysis. *Endoperidium* grey-brown, sometimes with a pinkish tinge, often covered in a fine, whitish, crystalline pruina. *Peristome* fibrillose, raised, usually distinctly delimited by a raised rim. *Columella* cylindric to clavate, often inconspicuous. *Mature gleba* dark brown.
Basidia not seen. *Capillitial hyphae* yellow-brown, thick-walled, with a narrow lumen, slender, 3–6 µm diam., finely tapered to subacute tips 1 µm or less in diam., walls smooth or finely encrusted. *Spores* globose, dark brown, 4.5–5.5 µm diam. excluding ornament, (5–) 5.5–6.3 µm diam. including ornament, verruculose, verruculae irregular, 0.3–0.5 µm high.

Habitat: British collections all in beech (*Fagus*) woodland on calcareous soil; elsewhere in Europe often in coniferous woodland (see Palmer 1968; Sunhede 1989).

Frequency and distribution: known in Britain mainly from south and south-east England; most collections from Surrey, others from Wiltshire and Gloucestershire. However, this species occurs throughout Europe where it is one of the commonest earthstars according to Sunhede (1989).

Other remarks: this species is similar to *G. fornicatum* in having a fornicate exoperidium in which the mycelial layer is cup-like, separated from the fibrous layer except at the tips of the rays. It differs in its smaller fruitbodies which have a distinctly delimited peristome and, at least in fresh specimens, a whitish-pruinose endoperidium. In addition, the spores of *G. quadrifidum* are larger and ornamented with coarser and more irregular verruculae. *Geastrum minimum* is similar in many respects to *G. quadrifidum* but its fruitbodies are not fornicate.

Four-Rayed Earthstar

Fig. 70. *Geastrum quadrifidum*. (Denmark, Rubjerg, 1988, Christensen).

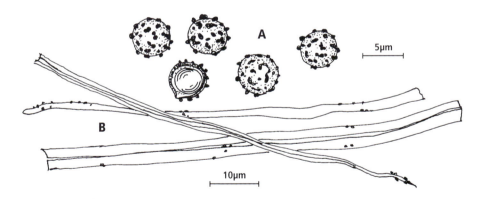

Fig. 71. *Geastrum quadrifidum*. (Surrey, Epsom, Oct. 1953, Orton). A, Spores; B, capillitial threads.

11. TINY EARTHSTAR

Map No. 34

Geastrum minimum Schwein. in *Schriften Naturf. Ges. Leipzig* 1: 58 (1822).
Geastrum marginatum Vittad., *Monogr. Lycoperd.*: 19 (1842, as *Geaster*).
Geastrum cesatii Rabenh. in *Bot. Zeitung (Berlin)* 9: 628 (1851, as *Geaster*).
Geastrum granulosum Fuckel in *Jahrb. Nassauischen Vereins Naturk.* 15: 41 (1860, as
Geaster).

Selected descriptions: Sunhede (1989: 256–281).

Selected illustrations: Dörfelt (1985: fig. 5/3, 42, 43); Ellis (1981: figs 1/13 & 3/6); Michael, Hennig & Kreisel (1986: fig 169); Sunhede (1989: figs 107–112, 114).

Diagnostic characters: peristome fimbriate; fruitbody small, 1.5–3 cm when expanded; mycelial layer encrusting debris; spore-sac 7–10 mm, stalked; spores 6–7.5 µm.

Fruitbody subglobose or depressed, hypogeous in development, encrusted with debris. Expanded fruitbody small, usually 15–30 mm across, splitting into 6–10 rays, non-hygroscopic, rays arched or somewhat recurved beneath the fruitbody, sometimes horizontal. *Mycelial layer* present, persistent, encrusting soil and debris. *Fibrous layer* pale grey-brown, rarely exposed. *Pseudoparenchymatous layer* whitish at first, becoming brown, splitting with age and when dry, sometimes forming a collar around the stalk. *Endoperidial body* 7–10 mm diam., stalked, subglobose, often with apophysis. *Stalk* c. 1 mm high, brownish. *Endoperidium* pale grey-brown, sometimes whitish-pruinose with crystalline matter. *Peristome* fibrillose, delimited by a weak groove. *Columella* whitish, cylindric or clavate, sometimes poorly delimited. *Mature gleba* brown.

Basidia not seen. *Capillitial hyphae* 3–6 µm diam., yellow-brown, thick-walled, usually with a narrow lumen, gradually tapered towards the tips and occasionally forked, irregularly encrusted, particularly towards the tips. *Spores* dark brown, globose, (4.8–) 5–6 (–6.2) µm diam. excluding ornament, (5.5–) 6–7.5 µm diam. including ornament, ornamented with irregular, rather coarse, sometimes coalescent verruculae 0.4–0.7 µm high, up to 2 µm across.

Habitat: on well-drained, calcareous soil. British material from near dunes, but it occurs in a range of habitats elsewhere in Europe. Its ecology is discussed in detail by Sunhede (1989).

Distribution and frequency: very rare in Britain and known with certainty from only a single collection: Norfolk, Holkham Gap, on mossy bank under *Pinus nigra* by open dunes, 6 Oct. 1958, Ellis (K, ex Palmer, Mycol. Herb. 2364). Material reported by Rea (1927) is misdetermined (Palmer 1968), and other reports from Britain are unconfirmed. The species is widespread in Europe.

Other remarks: The endoperidium, in fresh condition, is often covered with a whitish, crystalline pruina, but this may be lacking or lost with age. In addition, the spores are comparatively large and ornamented with coarse, irregular warts. *Geastrum coronatum* is similar in many respects, but has much larger fruitbodies and slightly smaller spores. *Geastrum quadrifidum* is also similar and differs most obviously in its fornicate exoperidium.

Fig. 72. *Geastrum minimum.* (Austria, Föhrenwald, 28 May 1983, Mrazek).

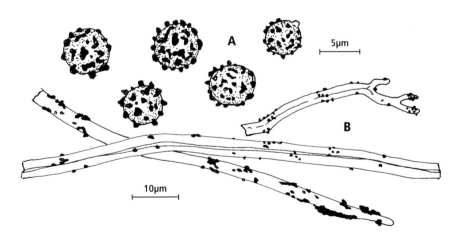

Fig. 73. *Geastrum minimum.* (Norfolk, Holkham, Oct. 1958, Ellis). A, Spores; B, capillitial threads.

12. ROSY EARTHSTAR

Map No. 35

Geastrum rufescens Pers.: Pers., *Synops. Meth. Fung.*: 134 (1801); *Neues Mag. Bot.* 1: 86 (1794).
Geastrum vulgatum Vittad., *Monogr Lycoperd.*: 20 (1842, as *Geaster*)
Geastrum schaefferi Vittad., *Monog. Lycoperd.*: 22 (1842, as *Geaster*).

Selected descriptions: Breitenbach & Kränzlin (1986: 503); Ellis (1981: 151, as *G. vulgatum*); Marchand (1976: p. 132); Sunhede (1989: 357–376).

Selected illustrations: Breitenbach & Kränzlin (1986: 503); Cetto (1988: 362); Dörfelt (1985: figs 54, 55); Ellis (1976: fig. 5); Ellis (1981: figs 2/7–8 & 3/14, as *G. vulgatum*); Gerhardt (1985: 209); Marchand (1976: pl. 364); Michael, Hennig & Kreisel (1986: fig 172)); Phillips (1981: 253); Ryman & Holmåsen (1984: p.602); Sunhede (1989: figs 155–163).

Diagnostic characters: peristome fimbriate; fruitbody large, 3–10 cm when expanded; mycelial layer encrusting debris; pseudoparenchymatous layer pinkish; spore-sac, 1–3.4 cm, pubescent, stalked; spores 4.2–5 μm.

Fruitbody partially hypogeous, subglobose to ovate or sometimes irregular in form, rarely slightly umbonate, encrusted with debris. Expanded fruitbody 26–90 mm diam., splitting to half way or more into 5–8 rays, non-hygroscopic, rays arched, spreading or sometimes recurved under the fruitbody. *Mycelial layer* present, persistent, encrusting soil and debris. *Fibrous layer* whitish or pale grey-brown. *Pseudoparenchymatous layer* whitish or pale fawn, becoming reddish brown, often splitting, hard when dry. *Endoperidial body* subglobose, sometimes with an apophysis, 11–30 mm diam., short-stalked. *Endoperidium* whitish at first, soon clay-coloured or brownish, often darker with age, minutely puberulent with protruding, hyaline or pale yellowish, usually tapered hyphae. *Peristome* fibrillose, broadly conical to flattened, not distinctly delimited. *Columella* clavate, whitish. *Mature gleba* dark brown, usually with a purplish tinge.

Basidia not seen. *Capillitial hyphae* 4–8 μm diam., yellow-brown, thick-walled, gradually tapered to a fine, subacute or rounded tip, distinctly encrusted over most of their length. *Spores* globose, dark yellow-brown to brown, verruculose, 3.8–4.3 μm diam. excluding ornament, 4.2–4.8 (–5) μm diam. including ornament, verruculae isolated, sometimes irregular, 0.2–0.3 (–0.4) μm high, 0.3–0.7 (–1) μm across.

Habitat: in woodlands or on heathlands, usually on calcareous, often sandy soil, sometimes in dunes. Mature in autumn.

Distribution and frequency: widespread but uncommon in England and Wales, and in Scotland known from only a few localities in the south. Occurs throughout Europe.

Other remarks: this species is similar in many respects to both *G. fimbriatum* and *G. coronatum*, though these lack the obvious pink or reddish colours which develop on the pseudoparenchymatous layer of *G. rufescens*. *Geastrum coronatum* otherwise differs in having a more distinctly stalked endoperidial body, with a non-puberulent endoperidium, less uniformly encrusted capillitial hyphae and larger, more coarsely ornamented spores.

Rosy Earthstar

Fig. 74. *Geastrum rufescens*. (Denmark, Ubriksdals, 21 Sept. 1986, Nitare).

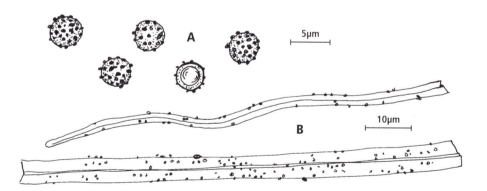

Fig. 75. *Geastrum rufescens*. (Kent, Polhill, Mar. 1980, Weightman). A, Spores; B, capillitial threads.

13. CROWNED EARTHSTAR

Map No. 36

Geastrum coronatum Pers.: Pers., *Synops. Meth. Fung.*: 132 (1801).
Geastrum limbatum Fr., *Syst. Mycol.* 3: 15 (1829, as *Geaster*).
Geastrum atratum F. Smarda in *Ceská Mykol.* 1: 74 (1947, as *Geaster*).

Selected descriptions: Ellis (1981); Sunhede (1989: 147–164).

Selected illustrations: Dörfelt (1985: figs 7/1, 34); Ellis (1981: figs 2/11,12 & 3/3);
Gerhardt (1985: 210); Michael, Hennig & Kreisel (1986 fig. 167); Phillips (1981: 253);
Sunhede (1989: figs 54–60)

Diagnostic characters: peristome fimbriate; fruitbody 3–10 cm, globose; mycelial
layer encrusting debris; pseudoparenchymatous layer not pinkish; spore-sac 1–3.5 cm,
with distinct stalk; endoperidium lacking erect hairs; spores 5–6.5 µm.

Fruitbody at first hypogeous, becoming epigeous, subglobose or somewhat
depressed,1.5–4 cm diam., encrusted with debris. Expanded fruitbody 4.5–8 (–10) cm
diam., exoperidium splitting into 7–12 rays, non-hygroscopic. *Mycelial layer* present,
encrusted with soil and debris, occasionally left as a cup-like structure in the ground
(Sunhede 1989). *Fibrous layer* when exposed greyish-white. *Pseudoparenchymatous
layer* fleshy, up to 5 mm thick, whitish or pale brown, becoming dark brown, drying to
leave irregular, net-like remnants. *Endoperidial body* 19–28 (–34) mm diam., stalked,
subglobose or depressed, commonly with a distinct apophysis. Stalk 2–4 mm high,
greyish brown. *Endoperidium* at first pale brown with a thin, felty layer of hyphae,
becoming almost smooth and darker, grey-brown to blackish. *Peristome* fibrillose,
circular or broadly ellipsoid, sometimes with a paler margin, distinctly delimited or not.
Columella whitish, cylindrical to clavate or irregular in form. *Mature gleba* dark brown,
often with purplish tinge.
   Basidia not seen. *Capillitial hyphae* pale yellowish brown, thick walled, with narrow,
discontinuous lumen, gradually tapered to the tips, partially finely encrusted, especially
towards the tips. *Spores* globose, 4.5–5.5 µm diam. excluding ornament, 5–6.2 (–6.5)
µm diam. including ornament, dark brown, verruculose, verruculae 0.3–0.6 µm high,
often partly coalescent, 0.5–1 (–1.5) µm across.

Habitat: various, usually on well-drained soil (calcareous or nutrient-rich *teste* Sunhede
1989), often in hedgerows. British collections often with hawthorn (*Crataegus*).

Distribution and frequency: widely distributed but rare in Britain. Known mostly from
eastern, central and south-east England, with a few localities inWales, and a few old
records from southern Scotland. Elsewhere in Europe mostly southern in distribution
(Sunhede 1989), with a few records from Denmark, Norway and Sweden.

Other remarks: *Geastrum coronatum* is closely related to *G. fimbriatum* and *G.
rufescens*. The former differs in having smaller fruitbodies with a usually sessile
endoperidial body. *Geastrum rufescens* differs in its smaller spores and minutely
puberulent endoperidial surface. *Geastrum minimum* is also similar in many respects to
G. coronatum, but has much smaller fruitbodies with paler endoperidium, a usually
distinctly delimited peristome, and slightly larger spores.

Crowned Earthstar

Fig. 76. *Geastrum coronatum*. (Denmark, Skanderborg, Lange).

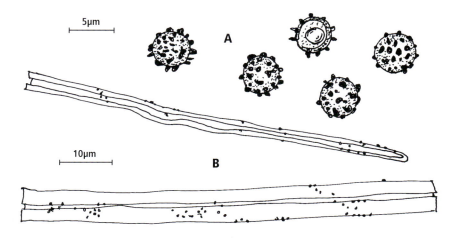

Fig. 77. *Geastrum coronatum*. (Devon, Torquay, Nov. 1954). A, Spores; B, capillitial threads.

British Puffballs, Earthstars and Stinkhorns

14. WEATHER EARTHSTAR

Map No. 37

Geastrum corollinum (Batsch) Hollós, *Magy. Gast.*: 57 (1903, as *Geaster*).
Lycoperdon corollinum Batsch, *Elenchus Fung.* 1: 151 (1783).
Lycoperdon recolligens With., *Bot. arr. Brit. Pl.* ed. 2, 3: 462 (1792).
Geastrum recolligens (With.) Desv. in *J. Bot. (Desvaux)* 2: 102 (1809).
Geastrum mammosum Chevall., *Fl. gén. env. Paris* 1: 359 (1826).

Selected descriptions: Ellis (1981, as *G. recolligens*); Sunhede (1989: 129–147).

Selected illustrations: Dörfelt (1985: fig. 32); Ellis (1981: fig. 1/4–5 & 3/10, as *G. recolligens*); Michael, Hennig & Kreisel (1986: fig. 170); Ryman & Holmåsen (1984: p.601); Sunhede (1989: figs 45–50).

Diagnostic characters: peristome fimbriate; fruitbody onion-shaped and umbonate, strongly hygroscopic; mycelial layer not encrusting debris, not splitting radially.

Fruitbody at first onion-shaped, umbonate, 1–2.5 cm across, 1–3 cm high. Expanded fruitbody 1.5–6 cm across, exoperidium splitting into 6–10 (–13) rays which are strongly hygroscopic, usually recurved under the fruitbody when damp, turning upwards. *Mycelial layer* present, pale brown, with inner whitish layer which is soon exposed and eventually flakes off in patches. *Fibrous layer*, when exposed, pale grey-brown, often with longitudinal striae. *Pseudoparenchymatous layer* persistent, 0.5–1.5 mm thick when fresh, greyish or pale brown, often dark brown with age, hard when dry. *Endoperidial body* sessile, subglobose, 10–18 (–24) mm diam. *Endoperidium* smooth or minutely pruinose, usually greyish or pale brown. *Peristome* fibrillose, usually distinctly delimited. *Columella* cylindric or clavate, whitish, extending to about half the height of the gleba. *Mature gleba* dark brown.
   Basidia not seen. *Capillitial hyphae* 5–9 µm diam., gradually tapered to subacute ends, yellow-brown, thick-walled, with narrow, discontinuous lumen, surface smooth or encrusted. *Spores* globose, with a single guttule, dark brown, verruculose, 3.6–4.5 µm diam. excluding ornament, 4–5 µm diam. including ornament, verruculae blunt, irregular, 0.2–0.5 µm high, 0.5–1 µm across.

Habitat: on well-drained, base-rich soil (Sunhede 1989), usually in deciduous woods or in hedgerows with deciduous shrubs, but also reported with *Juniperus* in Sweden.

Distribution and frequency: very rare in Britain, with few records since 1938. It is known mainly from Norfolk and Suffolk. Elsewhere in Europe it is widespread but rare (Sunhede 1989).

Other remarks: The first British collections were from Norfolk in 1782 (Ellis 1981), and further 18th century material was obtained from other localities in Norfolk and in Suffolk. The species was then uncollected until 1895 when discovered at Hillington, Norfolk by C. B. Plowright, who made collections in 1896 and 1897. It was also collected near Maidenhead, Berkshire in 1896. This century, it has been collected only from Wensleydale, Yorkshire in 1908 and, according to Ellis (1908), from Hellesdon, Norfolk, in 1938. The few records since then require confirmation. This species is similar to *G. floriforme*, but differs in having a comparatively well-delimited peristome, a mycelial layer which does not bind debris, and smaller spores.

Weather Earthstar

Fig. 78. *Geastrum corollinum.* (Sweden, Oland, Borgholm, Sunhede).

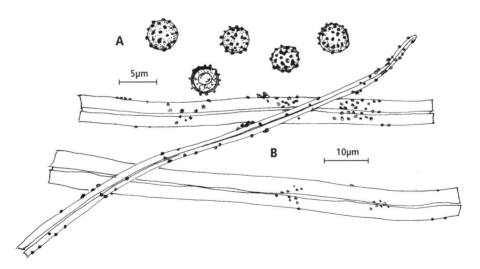

Fig. 79. *Geastrum corollinum.* (Norfolk, Hillington, Oct. 1895), Plowright). A, Spores; B, capillitial threads.

British Puffballs, Earthstars and Stinkhorns

15. COLLARED EARTHSTAR

Map No. 38

Geastrum triplex Jungh. in *Tijdschr. Natuurl. Gesch. Physiol.* 7: 287 (1840, as *Geaster*).
Geastrum michelianum W.G. Sm. in *Gard. Chron.* 1873: 608 (1873, as *Geaster*).
Geastrum cryptorhynchum Hazsl. in *Grevillea* 3: 162 (1874, as *Geaster*) *teste auct.*
Geastrum kalchbrenneri Hazsl. in *Verh. Zool.-Bot. Ges. Wien* 26: 220 (1876) *teste auct.*
Geastrum pillotii Roze in *Bull. Soc. Mycol. Fr.* 4: xxxvi (1888, as *Geaster*) *teste auct.*

Selected descriptions: Breitenbach & Kränzlin (1986: 502); Marchand (1976: p. 130); Sunhede (1989: 445–462).

Selected illustrations: Breitenbach & Kränzlin (1986: 502); Cetto (1988: 360); Dörfelt (1985: figs 6/5, 25, 69); Ellis (1976: fig. 6); Ellis (1981: figs 2/1–3 & 3/13); Gerhardt (1985: 208); Marchand (1976: pl. 363); Michael, Hennig & Kreisel (1986: fig 165); Phillips (1981: 253); Ryman & Holmåsen (1984: p.602); Sunhede (1989: figs 202–207).

Diagnostic characters: fruitbody large, fleshy; peristome fimbriate; not hygroscopic; mycelial layer without radial splitting, not encrusting debris; collar usually present; spore-sac large, smooth.

Fruitbody epigeous, distinctly umbonate and onion-shaped, with basal tuft of mycelium, not encrusted with debris. Expanded fruitbody 36–125 mm across, splitting into (4–) 5–7 rays; rays non-hygroscopic, arched, sometimes recurved under the fruitbody. *Mycelial layer* persistent, at first violaceous-pink (Palmer 1955), later buff to yellowish brown, developing paler, longitudinal splits large scales. *Fibrous layer* papery, pale buff or pale grey-brown. *Pseudoparenchymatous layer* fleshy, whitish to pale buff, often darker with age and when dry, commonly splitting at the margin of the exoperidial disc and becoming raised into a conspicuous collar, hard and rather brittle when dry. *Endoperidial body* sessile, subglobose, sometimes depressed, without an apophysis, 21–39 mm diam. *Endoperidium* smooth, fawn to pale buff or clay. *Peristome* fibrillose, broadly conical, usually fairly distinct, delimited by a faint circular depression. *Columella* whitish, narrowly clavate. *Mature gleba* dark brown to snuff-brown, without purple tint.
Basidia not seen. *Capillitial hyphae* 3–7 µm diam., pale yellow-brown to almost hyaline, gradually tapered to obtuse or subacute tips, thick-walled, distinctly encrusted over much of their length. *Spores* globose, yellow-brown to dark brown, verruculose, 3.5–4.2 µm diam. excluding ornament, 4.5–5.5 µm diam. including ornament, verruculae mostly isolated, blunt-cylindric, 0.4–0.6 µm high, 0.3–0.7 µm across.

Habitat: on well-drained, usually calcareous soil, on humus, leaf litter and compost, in deciduous woodland or in more open areas.

Distribution and frequency: one of the most common earthstars in Britain, and frequent throughout much of Europe. Especially abundant in eastern and southern England, but apparently less frequent northwards and uncommon in Scotland. It has also been reported from Ireland. This species is widely distributed in both temperate and subtropical areas.

108

Other remarks: examination of cotype material of *G. michelianum* (Castle Ashby, Nov. 1869, K) confirms the synonymy with *G. triplex* suggested by various authors such as Stanek (1958), Palmer (1968) and Dörfelt (1985). It is indeed surprising, as noted by Palmer (1968), that this species had not been distinguished in Britain prior to its description as *G. michelianus* by Smith; it had been previously misdetermined as *G. tunicatus*, *G. lageniforme* and others (see Palmer 1968). It is closely related to *G. lageniforme* and the differences between them are discussed below under that species.

Fig. 80. *Geastrum triplex*. (Surrey, Kew, Dec. 1990, Laessøe).

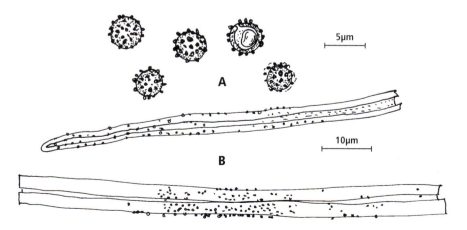

Fig. 81. *Geastrum triplex*. (Norfolk, Holkam, Oct. 1958, Palmer). A, Spores; B, capillitial threads.

British Puffballs, Earthstars and Stinkhorns

16. FLASK-SHAPED EARTHSTAR

Map No. 39

Geastrum lageniforme Vittad., *Monogr. Lycoperd.*: 16 (1842, as *Geaster*).

Selected descriptions: Sunhede (1989: 242–256).

Selected illustrations: Dörfelt (1985: fig. 40); Michael, Hennig & Kreisel (1986: fig 166); Sunhede (1989: figs 100–106).

Diagnostic characters: not hygroscopic; peristome fimbriate; mycelial layer not encrusting debris, with radial splits; no collar; spore-sac 1-2 cm, with erect, thick-walled hairs when young.

Fruitbody epigeous, umbonate, onion-shaped to lageniform, with a basal tuft of mycelium. *Expanded fruitbody* 24–60 mm diam, exoperidium splitting into 5–8 rays, non-hygroscopic, rays spreading, often recurved beneath the fruitbody. *Mycelial layer* dull yellowish brown or pale brown, often with paler, longitudinal splits. *Fibrous layer* whitish. *Pseudoparenchymatous layer* whitish at first, then brown and often date-brown, sometimes eventually peeling off. *Endoperidial body* 9–19 mm across, sessile, subglobose or depressed, without an apophysis. *Endoperidium* at first minutely puberulent with thick-walled, hyaline, protruding hyphae 25–65 x 5–12 µm, becoming smooth or often slightly scurfy, pale grey-brown to fawn. *Peristome* fimbriate, usually clearly delimited, bounded by a weak circular ridge. *Columella* not clear in material examined, columnar to club-shaped *teste* Sunhede (1989). *Mature gleba* dark brown, lacking purple tint.

Basidia not seen. *Capillitial hyphae* 5–13 µm diam., gradually tapered to subacute or rounded tips, pale yellow-brown, sometimes almost hyaline in places and towards the tips, thick-walled with lumen often comparatively wide, 2–6 µm diam., surface irregularly finely encrusted. *Spores* globose, yellow-brown to dark brown, verruculose, 3.4–4.5 µm diam. excluding ornament, 4.1–5.5 µm diam. including ornament, verruculae mostly isolated, low-cylindrical, simple or slightly irregular, blunt, 0.4–0.6 µm high, 0.3–0.7 µm across.

Habitat: not clearly stated in British collections; amongst litter. On dry, sandy soil (Sunhede 1989).

Distribution and frequency: very rare in Britain. Recorded with certainty only from Devon, Wiltshire, Herefordshire, Suffolk and Norfolk. It has apparently not been collected since 1953, and its present status remains uncertain. It is widely distributed in southern parts of Europe (Sunhede 1989).

Other remarks: It is closely related to *G. triplex*, and probably cannot be distinguished from it on spore characters alone. These species have been synonymized by some authors, such as Ponce de Leon (1968), but they nevertheless differ consistently in several respects. *Geastrum lageniforme* is smaller, with the endoperidial body rarely exceeding 20 mm in diam., and only rarely develops a pseudoparenchymatous collar. The endoperidium is finely puberulent, at least when young, with erect, thick-walled hyphae. Examination of neotype and authentic material of this species in K reveals the presence of thick-walled hyphae in the mycelial layer and fine, thick-walled protruding hyphae on the endoperidium of some fruitbodies. These characters are not mentioned by Sunhede (1989), yet seem to provide a useful means of distinguishing the species, at least when good, fresh material is available.

Fig. 82. *Geastrum lageniforme*. (Devon, Slapton,, 29 Aug. 1992, Roberts).

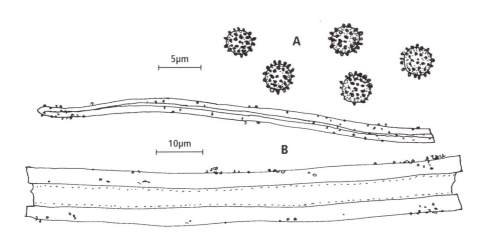

Fig. 83. *Geastrum lageniforme*. (Italy, Rome, Oct. 1846, neotype, K). A, Spores; B, capillitial threads.

British Puffballs, Earthstars and Stinkhorns

2. MYRIOSTOMA Desv.,
in *J. Bot. (Morot)* 2: 103 (1809).

Polystoma Gray, *Nat. arr. Brit. Pl.* 1: 586 (1821).

It differs from *Geastrum* in having a multistomatous endoperidium, and an endoperidial body borne on several stalks and containing several columellae. *Myriostoma* was referred to *Astraeaceae* by Dring (1973), but study of the arrangement and structure of the glebal tissue, and of the structure of the spore wall by Sunhede (1989) supports retention of the genus in *Geastraceae*. Monotype species: *M. coliforme* (With.: Pers.) Corda.

PEPPER POT
Map No. 40

Myriostoma coliforme (With.: Pers.) Corda in *Anleit. Stud. Mykol.*: lxxxi (1842).
Lycoperdon coliforme With., *Bot. arr. veg. Gr. Brit.* ed. 1, 2: 783 (1776).
Geastrum coliforme (With.) Pers., *Synops. Meth. Fung.*: 131 (1801).
Polystoma coliforme (With.) Gray, *Nat. arr. Brit. Pl.* 1: 586 (1821).
Myriostoma anglicum Desv. in *J. Bot. (Morot)* 2: 104 (1809).
Geastrum columnatum Lév. in *Ann. Sci. Nat. Bot.* sér. 3, 5: 161 (1846, as *Geaster*) *teste* Sunhede (1989).

Selected descriptions: Marchand (1976: p.120); Sunhede (1989: 468–486).

Selected illustrations: Cetto (1988: 368); Dörfelt (1985: figs 23, 63); Ellis (1981: fig. 1/15); Gerhardt (1985: 212); Marchand (1976: pl. 358); Michael, Hennig & Kreisel (1986: fig 158); Phillips (1981: 252); Sunhede (1989: figs 213–219).

Diagnostic characters: fruitbody *Geastrum*-like, large, 7–15 cm; spore-sac with numerous pores and stalks; 10–18 non-hygroscopic rays.

Fruitbodies epigeous, subglobose or depressed, not umbonate, with a basal tuft of mycelium. *Expanded fruitbody* 70–150 mm across, exoperidium splitting to about half way into 10–18 rays, non-hygroscopic, rays spreading, arched and often recurved at the tips. *Mycelial layer* persistent, pale brown or yellowish, usually cracking longitudinally and becoming scaly. *Fibrous layer* whitish to dull yellow-brown, sometimes with longitudinal splits. *Pseudoparenchymatous layer* fleshy when fresh, 3–5 mm thick, whitish to yellowish then brown, splitting and peeling away. *Endoperidial body* 27–57 mm diam., subglobose to depressed, supported by several (up to 17) stalks. *Stalks* visible only in dried specimens, usually flattened, 1–4 mm high. *Endoperidium* pale grey-brown, ornamented with numerous small warts and pits, glabrous, with several to numerous stomata. *Stomata* scattered, simple, slightly raised at first, soon flattened, circular or irregular in outline with slightly fimbriate margin, up to 3 mm diam. *Columellas* several, often branched, slender. *Mature gleba* dark brown.

Basidia not seen. *Capillitial hyphae* slender, mostly 4–5 μm diam., tapered to acute tips, occasionally forked, yellow-brown, thick-walled, with distinct but narrow lumen, surface smooth or slightly encrusted in places. *Spores* globose, yellow-brown, ornamented with irregular, curved, often branched and anastomosing ridges 0.7–1.2 μm high, 3.8–4.5 μm diam. excluding ornament, 6.2–7.2 μm diam. including ornament.

Pepper Pot

Habitat: recorded from roadside banks and hedgebanks, amongst nettles. Sunhede (1989) reports European material from well-drained, base-rich soil, especially near the sea.

Distribution and frequency: This unique and distinctive species was first described from Britain, although it has not been collected here since 1880, and must be considered extinct. However, it has been reported in recent years from the Channel Islands. It is widespread in north temperate and subtropical regions.

Other remarks: a unique species, easily recognized by the multi-stomatous endoperidium and multi-stalked endoperidial body. The prominent ridge- or wing-like ornament to the spores is also distinctive.

Fig. 84. *Myriostoma coliforme*. (Denmark, 14 Oct. 1987, Nitare).

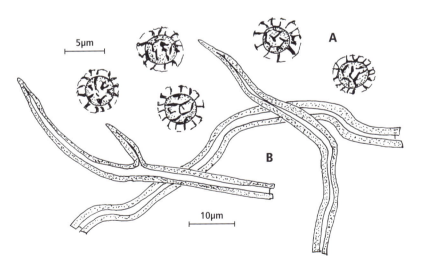

Fig. 85. *Myriostoma coliforme*. (unlabelled). A, Spores; B, capillitial threads.

2. LYCOPERDACEAE Chevall.
Fl. gén. env. Paris: 348 (1826).

Fruitbody epigeous at maturity, epigeous or hypogeous in development, sessile or raised on pseudostipe, globose, pyriform to pestle-shaped. *Peridium* 2–3 layered, outer layers either persistent or highly fugacious; *exoperidium* spinulose, granulose or smooth; *endoperidium* smooth, completely disintegrating to expose a mature gleba or typically opening through an apical pore or slit. *Subgleba* absent or present, massive or alveolate-poroid, *pseudocolumella* present or not. *Capillitium* of variable structure, occasionally only present as *paracapillitium*, with characteritic pores or slits or rarely non-poroid, rigid or fragile, hardly or highly branched, with or without true septa. *Basidia* clavate, with distinct sterigmata. *Basidiospores* almost hyaline to medium brown, globose to ovoid, smooth to verrucose, with or without attached sterigma remnant, less than 6 µm diam., often cyanophilous and dextrinoid. *Sphaerocysts* of thick-walled, dark coloured elements often present in exoperidium. On soil and humus, occasionally on wood. Type: *Lycoperdon* Tourn.: Pers.

The family is cosmopolitan but probably more abundant in the temperate zone and in dry, either hot or cool climates. All species appear to be saprophytes; some form fairy rings, others form rhizomorphs. The genus *Disciseda* Czern, unknown from Britain but with two species along the English Channel and North Sea coast, is charactezed by small bovistoid fruitbodies which turn over to reveal a basal, but functionally apical, opening to the endoperidium. The exoperidium is often retained as a disc (apical in development, but basal in mature fruitbodies), and the capillitium is of the *Lycoperdon*-type.

Another family of the Lycoperdales, *Mycenastraceae*, also occurs in continental Europe. This includes *Mycenastrum corium* (Guers.) Desv., superficially resembling a species of *Handkea* or *Calvatia* but with a very tough peridium.

Key to British genera

1. Sterile base (subgleba) absent or rudimentary 2
1. Sterile base prominent 3

    2. Fruitbody more than 10 cm in diam., sessile; capillitium of *Lycoperdon*-type .... **1. Calvatia**
    2. Fruitbody less than 6 cm diam; capillitium either of *Bovista*-, of intermediate or of *Lycoperdon*-type .... **5. Bovista**

3. Gleba separated from the sterile base by a well developed membrane (diaphragm); capillitium of *Lycoperdon*-type, non-poroid .... **2. Vascellum**
3. Diaphragm not developed; capillitium of various types, poroid or non-poroid 4

    4. Fruitbody turbinate with rooting base, and a wide opening; capillitium of *Bovista*-type .... **3. Bovistella**
    4. Not this combination of characters 5

5. Fruitbody large with disintegrating endoperidium; capillitium of *Lycoperdon*-type but with sinuous pores .... **4. Handkea**
5. Fruitbody small to medium, endoperidium not disintegrating; capillitium of various types 6

    6. Subgleba compact, poroid structure only visible under a lens; opens by slit or rarely by pore; capillitium of various types .... **5. Bovista**
    6. Subgleba poroid to naked eye; opens by a distinct pore; capillitium of *Lycoperdon*-type .... **6. Lycoperdon**

1. CALVATIA Fr.

Summ. Veg. Scand. 2: 442 (1849, nom. cons.).

Langermannia Rostk. in *Sturm, Deutschl. Fl.* 3: 3 (1839), (nom. rejic.).
Lasiosphaera Reichardt in *Fenzl's Reise Austr. Fregatte Novara* 1: 135 (1870).
  Additional synonyms: see Kreisel (1992).

Type species: *Lycoperdon craniiforme* Schwein.

GIANT PUFFBALL

Map No. 41

Calvatia gigantea (Batsch: Pers.) Lloyd, *Myc. Writ.* 1, Note 269: 166 (1904).
Lycoperdon giganteum Batsch, *Elench. Fung.*: 237 (1786).
Lycoperdon giganteum Batsch : Pers., *Synops. Meth. Fung.*: 140 (1801).
Bovista gigantea (Batsch: Pers.) Gray, *Nat. arr. Brit. Pl.* 1: 583 (1821).
Langermannia gigantea (Batsch: Pers.) Rostk. in *Sturm, Deutschl. Fl.* 3:: 23 (1839).
Lasiosphaera gigantea (Batsch: Pers.) Smarda in *Flora CSR*, B.1: 308 (1958).
Lycoperdon bovista Bull.: Pers., *Synops. Meth. Fung.*: 141 (1801).
Calvatia maxima (Schaeff.) Morgan in *Journ. Cinc. Soc. Nat. Hist.* 12: 166 (1890).

Selected descriptions: Zeller & Smith (1964: 167–169); Kreisel (1962: 121–123); Eckblad (1955: 32–33).

Selected illustrations: Marchand (1976: pl. 372); Dähncke & Dähncke (1979: 569); Michael & al. (1986: pl. 143); Ryman & Holmåsen (1984: 596); Phillips (1981: 247); Gerhardt (1985: 206); Bon (1987: 305); Lange & Hora (1965: 217, as *Lycoperdon giganteum*); Wakefield & Dennis (1981: pl. 109, f. 1, as *Calvatia gigantea*).

Diagnostic characters: large size; ball-like appearance; lack of evident subgleba.

Fruitbody depressed-globose, with a thick pseudorhiza, 20–50 (–70) cm diam, white, later yellowish to olive brown, very finely velvety to smooth. *Ectoperidium* thin, evanescent, white. *Endoperidium* thin, fragile, finally breaking away leaving exposed gleba. *Gleba* white, then yellow to dark olive-brown. *Subgleba* rudimentary, not alveolate. *Spore deposit* olive-brown.
  Capillitium of *Lycoperdon*-type with rounded pores and distant septa, sparsely branched, fragile, 2.5–9 μm diam., walls up to 1.5 μm, yellow brown. S*pores* globose to subglobose, 3.5–5.5 μm diam., asperulate, pale olive-brown, without sterigma remnants.

Habitat: in nutrient rich grasslands, parks, cultivated fields, compost heaps in gardens, hedgerows and woodland; nitrophilous.

Distribution and frequency: widespread and fairly common in Britain; appearing from summer. Temperate regions worldwide excluding S. America and western North America.

Other remarks: the giant fruitbody is among the largest known. Specimens of more than 20 kg in weight have been recorded and specimens over 4 kg are commonly recorded. Much sought after as an edible fungus. In continental Europe another five species of *Calvatia* occur. See notes under *Handkea*.

Giant Puffball

Fig. 86. *Calvatia gigantea* (Shropshire, Shrewsbury, Sept. 1989, Dickson).

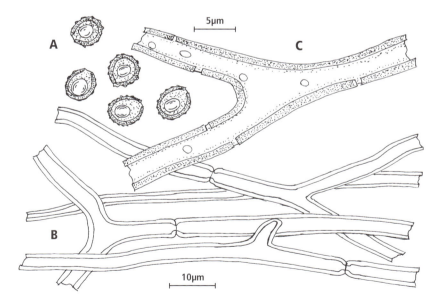

Fig. 87. *Calvatia gigantea* (Surrey, Kew, Apr. 1959, Brown). A, Spores; B, capillitial threads; C, portion of capillitial thread showing pores.

2. VASCELLUM Smarda in Pilát,
Flora CSR B.1: 304 (1958).

The genus *Vascellum* is characterised by the presence of a diaphragm, although this can vary in prominence. Some species have abundant true capillitium with or without pores, whilst the type species only has a rudimentary capillitium at the periphery of the gleba, the rest being hyaline, septate hyphae (paracapillitium). A columella is never developed. The subgleba is always distinctly poroid (alveolate). The genus is cosmopolitan, only lacking in arctic-subarctic climates. Twelve species are recognized by Kreisel (1993), two of which are known from Europe. *Type species*: *Vascellum depressum* (Bonord.) Smarda (= *V. pratense*).

MEADOW PUFFBALL

Map No. 42

Vascellum pratense (Pers.: Pers.) Kreisel in *Feddes Rep.* 64: 159 (1962).
Lycoperdon pratense Pers., *Tent. Disp. Meth. Fung.*: 7 (1797).
Lycoperdon pratense Pers.: Pers., *Synops. Meth. Fung.*: 142 (1801).
Lycoperdon hyemale Bull., *Champ. France*: 148 (1781).
Lycoperdon depressum Bonord. in *Bot. Zeitung* 15: 611 (1857).
Calvatia depressa (Bonord.) Z. Moravec in *Sydowia* 8: 284 (1954).
Vascellum depressum (Bonord.) Smarda, in *Flora CSR*, B.1: 305 (1958).

Selected descriptions: Kreisel (1962: 159–162); Smith (1974: 409–410); Monthoux (1982, SEM study); Monthoux & Röllin (1984: 196–199).

Selected illustrations: Breitenbach & Kränzlin (1986: pl. 521); Gerhardt (1985: 199); Michael & al. (1986: pl. 148); Cetto (1988: 347); Ryman & Holmåsen (1984: 597); Phillips (1981: 248); Bon (1987: 305); Lange & Hora (1965: 219, as *Lycoperdon depressum*); Wakefield & Dennis (1981: pl. 110, fig. 1. as *V. depressum*).

Diagnostic characters: small, flat-topped puffball; diaphragm separating powdery gleba from sterile base; true capillitium scarce; paracapillitium abundant.

Fruitbody in small groups, deep in substrate, more or less turbinate and flat-topped at maturity, 1.5–3.5 (–5) cm tall and 2–4.5 (–6) cm across. *Exoperidium* furfuraceous and spinose, white to ochraceous, spines up to 1.5 mm long, in groups of 6 or 8 with convergent tips, soon lost. *Endoperidium* smooth or with faint areolate pattern from worn-off exoperidium, white, in mature specimens grey-brown to grey. *Peristome* not developed but fruitbody open with an irregularly shaped aperture, up to 40 mm wide. *Gleba* white then grey-olive to grey-brown. *Diaphragm* distinct, skin-like. *Subgleba* comprising one-half to one-third of fruitbody, alveolate. *Spore mass* grey-olive to olive-brown.

Basidiospores globose or almost so, (2.5–) 3–4 (–4.5) μm, minutely verruculose, pale olive-brown, with inconspicuous sterigma remnant, rarely up to 2 μm long. *Capillitium* only present near endoperidial wall, non-poroid, 3–5 μm wide, similar to that found in species of *Lycoperdon*. *Paracapillitium* abundant throughout gleba, hyaline, septate, few branches, 2–6.5 μm wide.

Habitat: occurs in most types of grassland including lawns and open forests, but prefers drier habitats; tolerates the addition of artificial fertilizers and can be regarded as 'nitrophilous' (Arnolds 1982). Often associated with *Bovista plumbea*.

Meadow Puffball

Distribution and frequency: very common from early summer to late autumn throughout the British Isles and Eurasia. Almost cosmopolitan, but absent in lowland tropics.

Other remarks: Although diagnostic, the diaphragm can be difficult to locate and in such cases microscopical characters help in identification.

Fig. 88. *Vascellum pratense* (Austria, Schrems, 15 Oct. 1982, Mrazek).

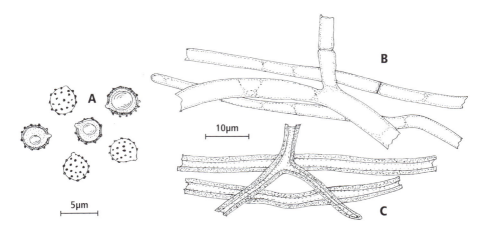

Fig. 89. *Vascellum pratense* (Surrey, Kew, Royal Botanic Gardens, 1993, Laessøe). A, Spores; B, paracapillitial threads; C, capillitial threads.

3. BOVISTELLA Morgan
in *J. Cincinnati Soc. Nat. Hist.* 14: 141 (1892).

Type species: *Bovistella radicata* (Dur. & Mont.) Pat. Only one species in Europe (Kreisel & Calonge 1993).

ROOTING BOVIST

Map No. 43

Bovistella radicata (Dur. & Mont.) Pat. in *Bull. Soc. mycol. Fr.* 15: 55 (1899).
Lycoperdon radicatum Dur. & Mont. in Dur., *Expl. Sci. Algérie* 1: 383 (1848).
   Additional synonyms: see Kreisel & Calonge (1993).

Selected descriptions: Reid (1953: 47–48); Kreisel and Calonge (1993: 19); Pilát (1937: 102–103); Schmitt (1978: 28–29).

Selected illustrations: Bull. Soc. Myc. France 42, Atlas pl. 14; Derbsch & Schmitt (1987: pl. 1, no. 166); Weber & Smith (1985: pl. 237); Schmitt (1978: pl.9 a–c); Reid (1953: 47; drawings); Kreisel & Calonge (1993: 19, drawing; 20, SEM microcharacters).

Diagnostic characters: resembles large *Vascellum pratense* or *Handkea utriformis*; *Bovista*-type capillitium; spores smooth, with long pedicels.

Fruitbodies in small groups or in fairy rings, very variable in size and shape, (2–) 3–7 (–25) cm diam., pyriform or more turbinate, with or without prominent pseudorhiza, normally firmly attached to substrate and long-lasting, opening with irregular, large, often star-like splitting of endoperidium. *Exoperidium* initially snow-white then cream to pale brown, furfuraceous and indistinctly spinulose; spines, especially on basal region, aggregated into pyramidal pegs. *Endoperidium* smooth, light grey to yellow-brown, darker at base, very old fruitbodies more grey and shiny and then very thin and papery. *Pseudodiaphragm* present, cup-shaped, felty. *Pseudocolumella* not present. *Gleba* initially yellow then deep ochre-brown to umber. *Subgleba* present and prominent, alveolate, occupying one third to half the fruitbody. *Spore deposit* olive-brown to umber.
   Capillitium of *Bovista*-type, branching, elastic, up to 7.5 (–14) µm broad at main branches, walls up to 1.5 µm thick, with regular to angular pores just visible on broader parts, golden brown to almost hyaline at tips. S*pores* 4.5–5 x 3.5–5 µm, globose to subglobose, appearing smooth to slightly asperulate (verruculose as seen by SEM), pedicels long, 2–15 µm, cylindrical, truncate.

Habitat: dry, more or less acid grasslands, roadside verges in open woodland, on well-drained, non-calcareous soils; appearing from May.

Distribution and frequency: known from only two sites in England and not collected in recent years; elsewhere known mainly from the warmer and continental parts of Europe, but nowhere common. Also known from northern Africa and North America, where said to be common in southern parts.

Other remarks: the genus was monographed by Kreisel & Calonge (1993) who recognized five species of which only the above is known from Europe.

Fig. 90. *Bovistella radicata* (USA, Ohio, Herb. Crossland).

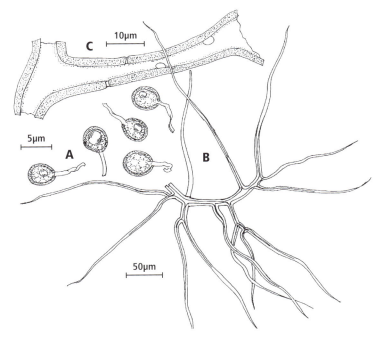

Fig. 91. *Bovistella radicata* (Surrey, Richmond, 1952, Walters). A, Spores; B, capillitial thread; C, portion of capillitial thread showing minute pores.

4. HANDKEA Kreisel

in *Nova Hedwigia* 48: 282 (1989).

Fruitbody mostly large, sub-globose, turbinate to pestle-shaped, splitting irregularly at maturity, upper part of endoperidium eventually falling away to expose powdery gleba; persists for long periods, even several years. *Exoperidium* variously ornamented with appressed scales or spines. *Endoperidium* smooth, not dextrinoid or very weakly so. *Subgleba* present and prominent, alveolate. *Pseudodiaphragm* often present. *Spore deposit* olive-brown to brown.

Capillitium fragile, branched, false-septate, similar to that of *Lycoperdon* and *Calvatia* but with slit-like (sinuous) pores, end segments long and tapering. *Basidia* as in the family; *basidiospores* more or less globose, from almost smooth to verrucose, with or without attached sterigmal remnants or in some cases with sterigmal remnants loose in spore mounts, acyanophilic. *Peridial elements* cyanophilic and dextrinoid, reaction depending on developmental stage. *Type species*: *Handkea utriformis* (Bull.: Pers.) Kreisel.

The genus is mainly characterized by the unique sinuous, slit-like pores in the capillitium. Previously, species of *Handkea* were mainly referred to the genus *Calvatia*, only represented in the British Isles by the Giant Puffball (*C. gigantea*), which lacks a distinct subgleba, a character that distinguishes a section within *Calvatia* sensu Kreisel. There are true septa in the capillitium of *Calvatia* species and Kreisel (1989, 1992) further stressed the presence of small rounded pores unlike the characteristic sinuous ones in *Handkea*. The endoperidium is dextrinoid in *Calvatia* and not, or only weakly so, in *Handkea* according to Kreisel, whilst Lange (1990) rejected this dichotomy. Demoulin (1993) accepted that the taxonomy of *Calvatia* is artificial but preferred not to change generic limits. Lange (1990) described and illustrated a range of subarctic and arctic species of *Handkea* and *Calvatia* (all as *Calvatia*) and also discussed generic limits, ecology and other aspects. He only accepted one genus, *Calvatia*, in the complex. Calonge & Martin (1990) also divided *Calvatia* sensu lato in another way than that followed here; *Langermannia* was recognized at genus level and *Handkea* treated as a synonym of *Calvatia*.

Calvatia sect. *Langermannia* (the Giant Puffball and some closely related species) is superficially very similar to the genus *Bovista* but is clearly separated by capillitium of *Lycoperdon*-type and by the disintegrating endoperidium (Kreisel 1992).

There are no species of *Calvatia* sect. *Calvatia* known from Britain but several occur on the continent. The most likely to be discovered in warmer parts of the British Isles is *C. fragilis* (Vittad.) Morgan and in colder areas possibly *C. candida* (Rostk.) Hollós. Some of the arctic-alpine taxa could also occur at high elevation.

Key to British species

1. Basidiospores smooth, without sterigmal remnants on spores or in mounts; fruitbodies as broad as tall or broader, stipe not clearly differentiated; in open grassy habitats
.... **1. Mosaic Puffball** (*H. utriformis*)

1. Basidiospores verrucose, with sterigmal remnants in mounts or attached to spores; fruitbody typically taller than broad, pestle-shaped, more rarely shaped as a short-stiped *Lycoperdon*-species; mostly in wooded habitats
.... **2. Pestle-shaped Puffball** (*H. excipuliformis*)

Handkea

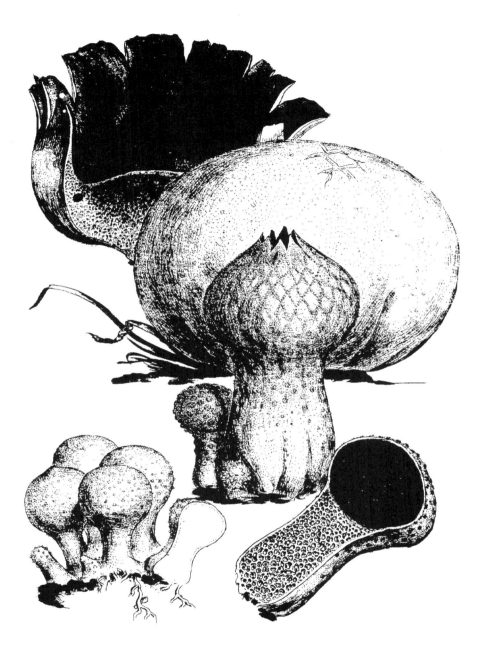

Fig. 92. From J. Sowerby's *Coloured Figures of English Fungi*, 1803, pl. 332, labelled 'Lycoperdon proteus' but representing species of *Calvatia, Handkea* and *Lycoperdon*.

1. MOSAIC PUFFBALL

Map No. 44

Handkea utriformis (Bull.: Pers.) Kreisel in *Nova Hedw.* 48: 288 (1989).
Lycoperdon utriforme Bull., *Champ. Fr.*: 153 (1791).
Lycoperdon utriforme Bull.: Pers., *Synops. Meth. Fung.*: 143 (1801).
Calvatia utriformis (Bull.: Pers.) Jaap in *Verhandl. Bot. Ver. Prov., Brandenb.* 59: 37 (1918).
Lycoperdon caelatum Bull., *Champ. Fr.*: 156 (1791).
Calvatia caelata (Bull.) Morgan in *J. Cincinnati Soc. Nat. Hist.* 12: 169 (1890).
Lycoperdon bovista Pers.: Pers., *Synops. Meth. Fung.*: 141 (1801) non *L. bovista* L.
Lycoperdon sinclairii Berk. ex Massee in *J. Roy. Microscop. Soc. London* 1887: 716 (1887).
   Further synonyms: see Kreisel (1989).

Selected descriptions: Kreisel (1962: 165–168); Eckblad (1955: 35–36); Lange (1990: 543–545).

Selected illustrations: Marchand (1976: pl. 374); Dähncke & Dähncke (1979: 568); Michael & al. (1986: pl. 144); Moser & Jülich VII: 1 & 5 (typical); Ryman & Holmåsen (1984: 595); Phillips (1981: 246); Bon (1987: 305); Lange & Hora (1965: 217, as *L. caelatum*); Wakefield & Dennis (1981: pl. 110, f. 4).

Diagnostic characters: large size; as broad as high; scaly surface; smooth spores and broad capillitium.

Fruitbody 5–15 cm broad, more or less turbinate, often depressed, base typically deeply sulcate, later cup-shaped with exposed gleba and very leathery. *Exoperidium* whitish then ochraceous to grey-brown, divided in polygonal plates, furfuraceous and composite spinulose, spines up to 4 mm, clusters up to 6 mm broad. *Endoperidium* lost at maturity. *Gleba* white then yellow through olivaceous to rich brown, loose and powdery. *Pseudodiaphragm* present and well marked. *Subgleba* persistent and solid, coarsely chambered (alveolate-poroid). *Spore deposit* brown.
   Basidiospores appearing smooth, 4–5 µm diam., globose to subglobose, 'apiculus' very indistinct, no sterigmal remnants present in slides. *Capillitium* fragile, sparsely branched, swollen at branching points, 2–15 µm broad, predominantly broad, ends wavy, taper to 2 µm, false septa present but rare, brown in maturity, pores slit-like. Cells in outer exoperidium up to 50 x 30 µm, in chains. Endoperidial hyphae dextrinoid [not acc. to Kreisel] and cyanophilous, of sub-parallel hyphae, 3–5 µm broad and non-septate.

Habitat: mostly unfertilized grasslands including dunes on acid to neutral soils (see Arnolds 1982).

Distribution and frequency: widely distributed and relatively common but rather local; widespread in the temperate zones, not extending into arctic-alpine climatic zones.

Other remarks: Kreisel (1989) recognized two varieties besides the type variety. Only the type variety occurs in western Europe. Lange (1990) considered the smooth spores, presence of pseudodiaphragm and the very broad, rarely septate capillitium as characters that could lead to a separation of *H.* (as *Calvatia*) *utriformis* at the generic level.

Mosaic Puffball

Fig. 93. *Handkea utriformis* (London, Putney, July 1992, Laessøe).

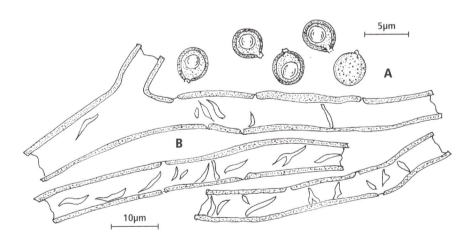

Fig. 94. *Handkea utriformis* (London, July 1991, Brown). A, Spores; B, capillitial threads with transverse slits.

2. PESTLE-SHAPED PUFFBALL

Map. No. 45

Handkea excipuliformis (Pers.: Pers.) Kreisel in *Nova Hedw.* 48: 283 (1989).
Lycoperdon excipuliforme Pers.: Pers., *Synops. Meth. Fung.* 1: 143 (1801).
Lycoperdon elatatum Massee in *J. Roy. Microscop. Soc. London* 1887: 710 (1887).
   Additional synonyms: see Kreisel (1962) and Palmer (1968).

Selected descriptions: Kreisel (1962: 168–172); Lange (1990: 540–543); Eckblad (1955: 37–38).

Selected illustrations: Marchand (1976: pl. 373); Dähncke & Dähncke (1979: 567); Michael & al. (1986: pl. 146); Moser & Jülich (1989: 4); Phillips (1981: 246); Bon (1987: 305); Lange & Hora (1965: 217); Wakefield & Dennis (1981: pl. 111, f. 2).

Diagnostic characters: large and pestle-shaped; smaller subglobose forms can be separated from species of *Lycoperdon* (esp. *L. molle*) on the strongly verrucose spores and sinuous pores in the capillitium.

Fruitbody 5–20 cm tall, 5–10 cm broad, pestle-shaped or rarely turbinate or pyriform, often sulcate from base to top of pseudostipe, overmature specimens cup-shaped or cylindrical. *Exoperidium* scurfy-furfuraceous, with fugacious composite spines, pale ochraceous to grey-brown. *Endoperidium* smooth, disappears in maturity. *Gleba* white, through yellow to chocolate-brown, powdery. *Pseudodiaphragm* indistinct or lacking. *Subgleba* firm and long lasting. *Spore deposit* brown.
   Basidiospores 4.5–5.5 µm diam. (6–7.5 µm with ornament), globose, verrucose, often with short sterigmal remnants, mounts often mixed with remnants of sterigmata. *Capillitium* fragile, sparsely branched, 2–5 (–10) µm broad, yellowish brown, pores small, slit-like. *Exoperidium* with fascicles of inflated cells. *Endoperidium* of dextrinoid, thickwalled hyphae.

Habitat: on humus, tolerates many soil types; mostly in woodland and plantations but also occasionally in grassland.

Distribution and frequency: widespread and common; occurs throughout most northern temperate regions with extensions to subarctic and subtropical climates.

Other remarks: Kreisel (1989) gave a very long list of synonyms and stated the interspecific variation to be very great but as yet unresolved. Kreisel (1962) recognized five different 'forms' mainly based on the shape of the fruitbodies.

Pestle-shaped Puffball

Fig. 95. *Handkea excipuliformis* (Surrey, Fetcham, Oct. 1992, Laessøe).

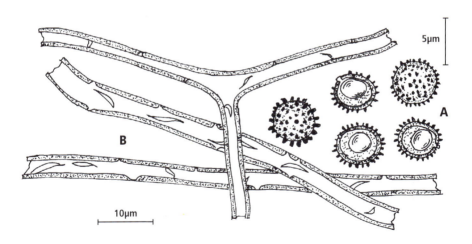

Fig. 96. *Handkea excipuliformis* (Surrey, Fetcham, Oct. 1992, Laessøe). A, Spores; B, capillitial threads.

5. BOVISTA Pers.: Pers.,

Synops. Meth. Fung.: 136 (1801) emend Kreisel in *Feddes Repert.* 69: 200 (1964).

Fruitbodies growing singly or in groups, tiny to medium sized, globose to pear-shaped with short pseudostipes, with or without rhizoids, fixed or loose (tumblers) at maturity; opening mostly irregular, slit-like but also in the form of a distinct raised operculum. *Exoperidium* mainly fugacious, flaking off and often completely lost at maturity or left as inconspicuous appressed squamules or pegs. *Endoperidium* smooth, papery, from dark brown to almost white, esp. in weathered specimens. *Gleba* initially white then of various olive-brown to purple-brown hues and powdery; pseudocolumella absent or rarely present. *Subgleba* developed or not, always of compact tissue. *Diaphragm* always absent. *Spore deposit* from olive to umber.

Basidiospores globose to ovoid, almost smooth to verruculose, with or without sterigmal remnant attached. *Basidia* as in family. *Capillitium* ranging from *Bovista*-type via intermediates to *Lycoperdon*-type, either non-poroid or with various amounts of differently sized pores. *Paracapillitium* mostly absent at maturity. *Exoperidium* variously composed, often (always?) two-layered. *Type species*: *Bovista plumbea* Pers.: Pers.

There is still no consensus concerning the delimitation of *Bovista* and *Lycoperdon*. Two opposite views are expressed in, for example, Kreisel (1967) and Lange (1987). Here the broad definition imployed by Kreisel (1967) in his monograph has been followed for convenience. Concerning species separation in *Bovista* several schools also exist. Here a fairly broad species concept has been adhered to.

Key to British and potentially British species

1. Spores without sterigmal remnants; capillitium of *Lycoperdon*-type or of intermediate type 2
1. Spores with sterigmal remnants, capillitium of *Bovista*-type or of intermediate type 3

   2. Subgleba well developed; spores weakly asperulate; in calcareous habitats, in the open or in forests .... **1. Deceiving Bovist** (*B. aestivalis*)
   2. Subgleba not present or rudimentary; spores distinctly verrucose; on acidic to neutral humose soils in open habitats .... **2. Dwarf Bovist** (*B. dermoxantha*)

3. Subgleba well developed; capillitium of *Bovista*-type; associated with moss in calcareous fens .... **3. Fen Bovist** (*B. paludosa*)
3. Subgleba absent; capillitium of *Bovista*- or intermediate type; normally occurring in drier habitats 4

   4. Capillitium of intermediate type; fruitbodies tiny, less than 15 mm diam., opening by raised delimited operculum; in calcareous dunes .... **4. Least Bovist** (*B. limosa*)
   4. Capillitium of *Bovista*-type; fruitbodies 10–60 mm diam., opening by slit or irregularly, without raised operculum; in dunes and other types of grassland 5

5. Fruitbody small, less than 20 mm diam., attached at maturity; capillitium with thinner parts poroid; in dry sandy habitats .... (*B. tomentosa*)
5. Fruitbody 10–60 mm diam., loose at maturity; capillitium non-poroid; in grassland 6

   6. Spores with curved sterigmal remnants; endoperidium brown ...... (*B. graveolens*)
   6. Spores with straight sterigmal remnants 7

7. Endoperidium dark brown, shiny; fruitbody 30–60 mm diam., spores globose with sterigmal remnants 4–9 (–13) µm long .... **5. Brown Bovist** (*B. nigrescens*)
7. Endoperidium mostly lead-grey, mat; fruitbody 10–25 (–55) mm diam.; spores subglobose to ovoid with sterigmal remnants (5–) 8–14 (–18) µm long
.... **6. Lead-grey Bovist** (*B. plumbea*)

1. DECEIVING BOVIST

Map. No. 46

Bovista aestivalis (Bonord.) Demoulin in *Beih. Sydowia*. 8: 143 (1979).
Lycoperdon aestivale Bonord., *Handb. Allg. Mykol.*: 251 (1851).
Lycoperdon cepaeforme Bull., *Champ. Fr.*: 156 (1791).
Lycoperdon furfuracea Schaeff., *Icon. fung. Bav.* 4: 131 (1774), non *Bovista furfuracea*
   (Gmelin : Pers.) Pers.
Lycoperdon polymorphum Vittad., *Monogr. Lycop.*: 183 (1843), nom. rej. (Art. 63, ICBN).
Lycoperdon coloratum Peck in *Ann. Rep.New York State Mus. Nat. Hist.* 29: 46 (1878).
Bovista colorata (Peck) Kreisel in *Feddes Repert.* 69: 201 (1964).
Bovista pusilliformis (Kreisel) Kreisel in *Feddes Repert.* 69: 202 (1964).

Selected descriptions: Calonge (1992: 103–104); Demoulin (1968: 55, as *B. polymorpha*);
Kreisel (1967: 108–115, as *B. polymorpha*).

Selected illustrations: Bresadola (1932: pl. 1140, 1, as *L. furfuracea*, very young);
Michael & al. (1986: pl. 147, as *B. polymorpha*); Vittadini (1842: pl. 2, fig. 7 (as
'VIII'!)).

Diagnostic characters: resembles species of *Lycoperdon* but with compact subgleba;
peridium often tinged reddish basally; capillitium of intermediate type; spores only with
faint ornament and sterigmal remnant not prominent.

Fruitbody 10–40 mm diam., globose to pear-shaped, with rhizoid, opening irregular,
4–5 mm wide. *Exoperidium* smooth or granulose, flaky, white, fugacious, when dry
ochraceous brown. *Endoperidium* thin, grey-brown. *Gleba* olive- to umber-brown.
Subgleba present or almost absent in small fruitbodies, compact, cellular structure only
visible with handlens. *Spore deposit* olive- to light umber-brown.
   Basidiospores 3.5–4.5 (–5) µm, globose, asperulate, lacking sterigmal remnant.
Capillitium of intermediate type, main branches 4.5–12 µm broad, with abundant,
minute, simple pores.

Habitat: on calcareous, mainly sandy soils, in dunes, downland and dry woodland.

Distribution and frequency: widespread in Britain, probably fairly common, occurring
particularly near the coast. Cosmopolitan; very common in the Mediterranean region
(Calonge 1992).

Other remarks: a very polymorphic and taxonomically difficult species. Calonge &
Demoulin (1975), Calonge (1992) and Demoulin (1979) discussed the problems. A
broad circumscription has been employed here. Collections from wooded habitats
like Mickleham Downs would probably belong with *B. pusilliformis* if a narrow
species concept was used. The species is often overlooked, having small and fairly
featureless fruitbodies.

Deceiving Bovist

Fig. 97. *Bovista aestivalis* (Denmark, Jutland, 29 Sept. 1985, Jeppson).

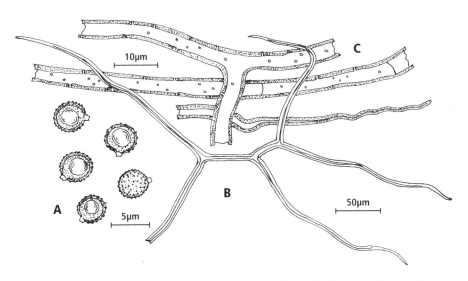

Fig. 98. *Bovista aestivalis* (Licolnshire, Aug. 1910). A, Spores; B, capillitial threads; C, detail of capillitial threads.

131

2. DWARF BOVIST

Map. No. 47

Bovista dermoxantha (Vittad.) De Toni in Sacc., *Syll. Fung.* 7: 100 (1888).
Lycoperdon dermoxanthum Vittad., *Monogr. Lycoperd.* 178 (1843).
Bovista pusilla (Batsch : Pers.) Pers., *Synops. Meth. Fung.*: 138 (1801) sensu Kreisel
  (1967) non sensu Persoon
Lycoperdon ericetorum Pers., in *J. Bot. (Desvaux)* 2: 17 (1809, nom. rej. ICBN Art. 63).

Selected descriptions: Calonge (1992: 110–111); Demoulin (1968: 55–57); Kreisel (1967: 63–71).

Selected illustrations: Gerhardt (1985: 204, as *B. pusilla*).

Diagnostic characters: small fruitbodies with subgleba absent or little developed; capillitium of *Lycoperdon*-type; spores distinctly ornamented.

Fruitbodies solitary or in small groups, (7–) 10–30 (–40) mm diam, globose to pear-shaped, typically with prominent, sand-encrusted rhizoid; opening round to irregularly split, 2–10 mm wide. *Exoperidium* smooth, white, later breaking up into small appressed squamules or tiny pegs. *Endoperidium* papery, grey-brown or occasionally of other hues. *Gleba* light to dark olive-brown. *Subgleba* absent or rarely present as trace of compact tissue. *Spore deposit* light olive-brown.
  Basidiospores 3.5–4.5 (–5.5) μm, globose, verrucose. *Capillitium* of *Lycoperdon*-type, branches up to 5.5 (–8) μm broad, with abundant pores of varying sizes.

Habitat: on humus in sun-exposed, dry, acid grasslands extending to margins of woodland and road verges.

Distribution and frequency: uncommon but widespread; overlooked and easily confused. Probably cosmopolitan but taxonomy not well understood.

Other remarks: very close to *B. aestivalis* but differs in the more coarsely ornamented spores and in the capillitium being solely of *Lycoperdon*-type without any intermediate elements. The species has mainly been recorded under the name *B. pusilla* but the type, a plate by Batsch, clearly depicts *B. limosa*. The best action seems to be to follow Ortega & Buendia (1989), who concluded that the preferred name should be *B. dermoxantha*. The name *B. pusilla* should be considered a *nomen ambiguum*. The loss of this well-known name could be viewed as disadvantageous but comparison of illustrations and herbarium material under this name shows widespread confusion in application of the name. All records of *B. pusilla* should be reviewed carefully. For a discussion of ecological requirements see Arnolds (1982, as *B. pusilla*).

Dwarf Bovist

Fig. 99. *Bovista dermoxantha* (Sweden, Västergötland, 17 Aug. 1985, Jeppson).

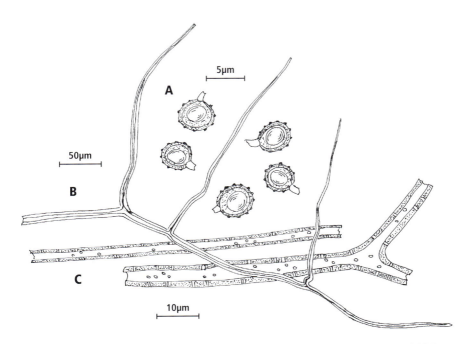

Fig. 100. *Bovista dermoxantha* (Lancashire, Ainsdale, June 1953, Palmer). A, Spores; B, capillitial thread; C, detail of capillitial threads.

3. FEN BOVIST

Map No. 48

Bovista paludosa Lév. in *Ann. Sci. Nat., Bot.* sér. 3, 5: 163 (1846).
Bovistella paludosa (Lév.) Pat. in Lloyd, *Mycol. Writ.* 1: 88 (1902).

Selected descriptions: Calonge (1992: 109); Favre (1937: 293–296); Kreisel (1967: 130–134).

Selected illustrations: Anonymous (1992: pl. 15); Poelt & Jahn (1963: pl. 175, middle, right); Rea (1909: pl. 8); Ryman & Holmåsen (1984: 593).

Diagnostic characters: habitat in fens; compact subgleba well developed; capillitium of *Bovista*-type.

Fruitbodies solitary or in small groups, 13–60 mm high with stipe part up to 35 mm long, more or less pear-shaped or rarely almost globose, without prominent rhizoids, opening with apical slit that later becomes irregularly torn. *Exoperidium* smooth, thick, white, later present as thin, scattered, grey-white, appressed scales. *Endoperidium* papery, yellow-brown, reddish brown to bronzy blackish brown. *Gleba* olive to olive-brown. *Subgleba* compact, prominent, olive- to grey-brown, grading into gleba. *Spore deposit* olive-brown.

Basidiospores 3.5–5.5 µm diam., globose to subglobose, asperulate, with sterigmal remnant 6.5–15 µm long, straight, truncate to pointed. *Capillitium* of *Bovista*-type, with short main branches to 6–12 µm broad, non-poroid, rarely septate. *Paracapillitium* not present in properly matured fruitbodies.

Habitat: amongst mosses in calcareous marshes (fens).

Distribution and frequency: Norfolk and Yorkshire fens and marshes; very rare but easily overlooked; northern hemisphere, possibly disjunct.

Other remarks: Gibbs (1908) and later Rea (1909) reported the first British collection (from Yorkshire) based on a determination by Lloyd. Subsequently, it has been recorded only by E.A. Ellis from the Norfolk fens, where it grew with the moss *Aulocomnium palustre*. These important fen sites should be reinvestigated and conservation measures, where necessary, should be considered. *Lycoperdon caudatum* is said to occur with *B. paludosa* in some sites in continental Europe (Kreisel 1976).

Fen Bovist

Fig. 101. *Bovista paludosa* (Denmark, Torne, 7 Sept. 1985, Nitare).

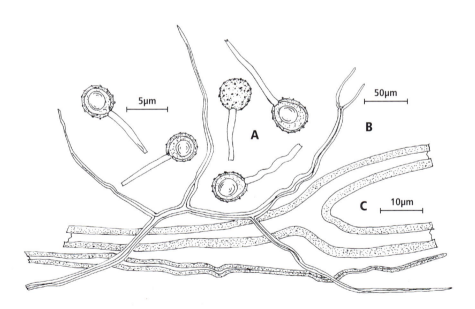

Fig. 102. *Bovista paludosa* (Norfolk, Sept. 1944, Ellis). A, Spores; B, capillitial thread; C, detail of capillitial threads.

4. LEAST BOVIST

Map No. 49

Bovista limosa Rostr. in *Meddel. Grønland.* 18: 52 (1894).
Lycoperdon limosum (Rostr.) Rauschert in *Z. Pilzk.* 25: 52 (1959).
  Misapplication: *Bovistella echinella* (Pat.) Lloyd sensu Lloyd & auct. non sensu Pat.

Selected descriptions: Calonge & Demoulin (1975: 263); Demoulin (1968: 57); Kreisel (1967: 116–120); Lange (1987: 268–270); Monthoux & Röllin (1976: 248–251).

Selected illustrations: Kreisel (1967: fig. 19a & 27c, 60 b&w); Lloyd (1910: 452 fig. 270 b&w, as *Bovistella echinella*); Lohwag (1933: pl.3, fig. 1a–b, b&w, *B. echinella*).

Diagnostic characters: tiny species; operculum often raised like the peristome in species of earthstar and stalk puffball; no subgleba; capillitium of intermediate type.

Fruitbody 5–15 mm diam., globose, without distinct rhizoids but relatively firmly attached to substratum. Opening small, to 3 mm diam., with raised margin as in *Geastrum*, either smooth, with rounded lobes or dentate, occasionally surrounded by a circular depression. *Exoperidium* smooth then becoming felty and typically covered in irregular polygons, with or without a small tuft in the middle, eventually either with small pointed tufts, sometimes stellate in appearance, or with endoperidium completely exposed. *Endoperidium* very thin, papery, smooth, matt or somewhat shiny, reddish to umber-brown. *Gleba* olive-brown to umber-brown. *Subgleba* absent. *Spore deposit* olive- to light umber-brown.
  Basidiospores (3.5–) 4–5.5 (–6) µm, globose, dispersed-asperulate, with sterigmal remnant 3–10 µm or broken off, mostly straight, cylindrical to somewhat pointed. *Capillitium* of intermediate type, with discrete elements lacking or very rare in the central part of the gleba; branching varies somewhat with position within the gleba, normally dichotomous, of *Lycoperdon*-type near the endoperidium, ends of threads typically somewhat spiralling, main branches mostly less than 5 µm broad but occasionally up to 7 (–10) µm broad, wall thickness to 1.5 µm; pores absent, true septa absent, false septa common.

Habitat: on sun-exposed sites with dry, calcareous, sandy soils, with sparse lichen and moss cover (e.g. *Tortula*); in the British Isles only from dunes but should be looked for in other dry, open habitats.

Distribution and frequency: Welsh and Lancashire dunes; rare, but easily overlooked. It is said to favour continental climates and to occur in similar situations as *B. tomentosa* (Lange 1987) and in dunes in similar situations as *Tulostoma brumale*. See Runge & Gröger (1990) for a detailed account of European records and habitats. Also known from North America.

Other remarks: occupies an intermediate position between *Lycoperdon* and *Bovista*, according to Lange (1987). The operculum is a unique feature, but not enough to warrant a separate genus (Lange 1987). The species was described from high-arctic Greenland and was, until recently, hardly known in Europe where, presumably, it had been overlooked due to size and habitat. Palmer (1968) reported the first collection from the British Isles. See also *Bovista dermoxantha*.

Fig. 103. *Bovista limosa* (Dyfed, Ynyslas, Rotheroe).

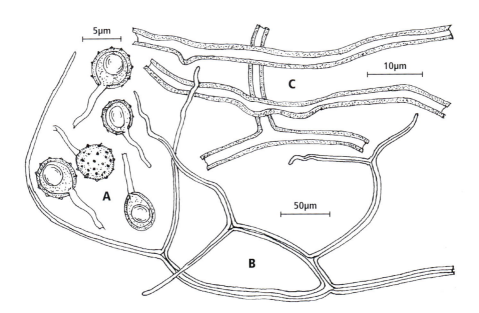

Fig. 104. *Bovista limosa* (Dyfed, Pembrey, Oct. 1991, Jones). A, Spores; B, capillitial thread; C, transitional capillitial thread, with small pores.

5. BROWN BOVIST

Map No. 50

Bovista nigrescens Pers.: Pers., *Synops. Meth. Fung.* 136 (1801); Pers. in *Roem., Neues Mag. Bot.* 1: 86 (1794).

Selected descriptions: Demoulin (1968: 60); Kreisel (1967: 151–157); Lange (1987: 265–267, peridium structure only).

Selected illustrations: Cetto (1988: 345); Gerhardt (1985: 203); Lange & Hora (1965: 219); Phillips (1981: 249).

Diagnostic characters: largest bovist; exoperidium quickly lost exposing brown endoperidium; operculum fairly wide; spores with straight sterigmal remnants.

Fruitbodies solitary or in small groups, 30–60 mm diam., globose, with irregular opening, starting as a slit, to 40 mm wide, no rhizoids; at maturity free from the substratum. *Exoperidium* white, smooth, soon splitting into appressed squamules and eventually lost completely. *Endoperidium* strong, smooth, shades of dark brown, more or less shiny. *Gleba* olive-brown to deep purple-brown. *Subgleba* not present. *Spore deposit* dark umber to purplish chestnut.

Basidiospores 4–6 µm diam., globose to subglobose, asperulate-verrucose; sterigmal remnant truncate, 4–9 (–13) µm long. *Capillitium* of *Bovista*-type, with main branches up to 22 (–32) µm broad, non-septate, non-poroid, tapering to fine points. *Paracapillitium* not present.

Habitat: mainly on humose, weakly acidic sand in a wide range of grassy habitats, like sheep walks, but also various types of woodland; tolerates and thrives with fertilization.

Distribution and frequency: widespread and common; increases towards the north and with altitude; mainly from June to September but fruitbodies can persist for very long periods. Common all over Europe especially at higher altitudes and in atlantic-subatlantic climates. Replaced by a close relative in North America.

Other remarks: the very similar *B. graveolens* should be looked for in Britain. The spores have a curved sterigmal remnant. It prefers cultivated habitats according to Kreisel (1967). The ecology was discussed in Arnolds (1982). Adverse conditions or picking of immature fruitbodies of *B. nigrescens* can lead to an abnormal maturation process making it difficult to identity the resulting specimens.

Brown Bovist

Fig. 105. *Bovista nigrescens* (Shropshire, Preston Montfort, Oct. 1978, Dickson).

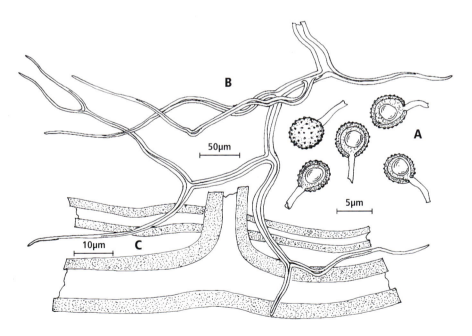

Fig. 106. *Bovista nigrescens* (Argyll, Kintyre, Sept. 1980, Dennis). A, Spores; B, capillitial thread; C, Detail of capillitial thread.

6. LEAD-GREY BOVIST

Map. No. 51

Bovista plumbea Pers.: Pers., *Synops. Meth. Fung.* 137 (1801); Pers., *Obs. myc.* 1:5 (1796).

Selected descriptions: Demoulin (1968: 59); Monthoux & Röllin (1976: 253 & 255); Kreisel (1967: 160–167).

Selected illustrations: Marchand (1976: pl. 365); Phillips (1981: 249); Michael & al. (1986: pl. 145); Bon (1987: 305); Lange & Hora (1965: 219); Wakefield & Dennis (1981: pl. 111, fig. 6); Gerhardt (1985: 203).

Diagnostic characters: lead grey endoperidium; no subgleba; spores with long sterigmal remnant; subglobose to ovoid, very faintly ornamented.

Fruitbodies in small groups, 10–25 (–55) mm diam., globose, loosening at maturity, opening slit-like to irregular, to 9 (–20) mm across. *Exoperidium* white, drying greyish white and loosening like the shell of a boiled egg. *Endoperidium* papery, typically lead grey and matt. *Gleba* olive to umber. *Spore deposit* olive-brown, sepia to umber.

Basidiospores 4.5–6.5 x 4–5.5 µm, subglobose to ovoid, almost smooth to asperulate, often with encrusting matter attached, sterigmal remnant (5–) 8–14 (–18) µm, cylindrical or attenuated, light olive-brown. *Capillitium* of *Bovista*-type, with main branches up to 15 (–35) µm broad, tapering to acute tips, non-poroid, non-septate. *Paracapillitium* absent in mature specimens.

Habitat: on humus in many types of grassland, including fertilized lawns.

Distribution and frequency: widespread and very common, decreasing towards the north and at higher altitudes. Cosmopolitan except lowland, wet tropics, although probably introduced to ruderal pastures in the tropics together with *Vascellum pratense* (Demoulin & Dring 1975).

Other remarks: the basidiospores are often referred to as smooth but a distinct, although very low, ornamentation can be seen, even with low-power objectives. For discussions on ecological requirements see Arnolds (1982) and Kreisel (1957, 1973).

Lead-grey Bovist

Fig. 107. *Bovista plumbea* (Surrey, Richmond, July 1991, Laessøe).

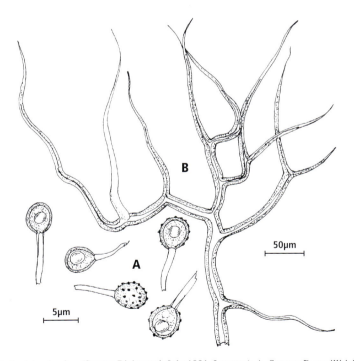

Fig. 108. *Bovista plumbea* (Surrey, Richmond, July 1991, Laessøe). A, Spores; B, capillitial thread.

6. LYCOPERDON Tournef. ex L.: Pers.
Synops. Meth. Fung.: 140 (1801).

Fruitbodies solitary or fasciculate, pear-shaped to pestle-shaped, with delimited round apical opening at maturity, pale ochraceous, pinkish, grey-brown to dark brown. *Exoperidium* scurfy to spiny or a mixture of such elements, rarely veiled, persistent or fugacious and then in some species leaving an areolate pattern, spines cream to dark brown. *Endoperidium* papery, mostly concealed by exoperidium. *Gleba* with or without distinct *pseudocolumella*, olive, chocolate-brown or purplish brown at maturity. *Subgleba* always well developed, poroid (alveolate) to the naked eye, mostly spongy. *Diaphragm* not present. *Spore deposit* of olive-brown to chocolate-brown hues.

    Basidiospores globose, less than 5 (–6) μm diam, smooth to verrucose with varying density of warts or spines. *Capillitium* of *Lycoperdon*-type, either elastic or fragile, with varying wall thickness, mostly poroid. *Paracapillitium* present in varying amounts. *Exoperidium* with more or less characteristic spherocysts.

    Mostly on well-drained soils in forests and open grassy habitats; humicolous, rarely lignicolous. Throughout the British Isles; cosmopolitan. *Type species*: *Lycoperdon perlatum* Pers.: Pers.

Features from the exoperidium are especially important. It is very easy to confuse those species of *Bovista* which have a well-developed subgleba with species of *Lycoperdon*. Species from the latter genus always have a distinctly 'chambered' subgleba, visible to the naked eye.

    Descriptions for the *Lycoperdon* entries have been improved by consulting the excellent annotated key to the Swedish species (Jeppson 1984) and likewise the important contributions made by Kreisel (1973) and Demoulin (1972a,b, 1978, 1983).

Key to British and some potentially British species.

1. On wood, with prominent white rhizoids, gasterocarp smooth at maturity, subgleba white; spores smooth, capillitium without pores ..... **1. Stump Puffball** (*L. pyriforme*)
1. On soil and humus, rarely on wood and then with prominent exoperidal ornamentation at maturity; subgleba not white at maturity; spores ornamented, capillitium mostly poroid. 2

    2. Fruitbody with white or pinkish flaky veil, underlying exoperidium with delicate spines; in calcareous woodland .... **2. Flaky Puffball** (*L. mammiforme*)
    2. Fruitbody unveiled; exoperidium variable; habitat variable 3

3. Spores with long, attached sterigmal remnant; fruitbody with long, convergent fugaceous spines leaving faint areolate pattern; in moist, wooded or open, calcareous localities .... **3. Pedicelled Puffball** (*L. caudatum*)
3. Spores without attached sterigmal remnants 4

    4. Mature gasterocarps with areolate pattern from shed exoperidial spines 5
    4. Mature gasterocarps without such areolate pattern 8

5. Exoperidial spines 3–6 mm long, dark brown; spores with prominent ornament; in ancient, nutrient-rich woodlands .... **4. Hedgehog Puffball** (*L. echinatum*)
5. Exoperidial spines shorter, of varying colours; spores with finer ornament 6

6. Exoperidium with fugacious, thick, solitary, conical spines surrounded by a persistent circle of discrete, low warts; very common in all types of woodland, occasionally on wood .... **5. Common Puffball** (*L. perlatum*)

6. Exoperidium with different pattern, often with spines aggregated in groups with convergent tips 7

7. Exoperidial spines pale brown, leaving a faint areolate pattern; spores 3.3–3.6 µm diam., almost smooth; woodland, boreal zone .... (*L. norvegicum*)

7. Exoperidial spines dark brown, leaving prominent areolate pattern; spores 3.8–4.4 µm, asperulate; in acid, open or wooded habitats; common .... **6. Blackish Puffball** (*L. nigrescens*)

8. Exoperidium mostly scurfy-granulose to slightly spiny at lower parts, fruitbodies often small and greyish; spores not mixed with sterigmal remnants; common in open sandy, mostly calcareous habitats .... **7. Grassland Puffball** (*L. lividum*)

8. Exoperidium with clearly differentiated spines 9

9. Spore preparations with large amount of detached sterigmal remnants; spores mainly coarsely ornamented 10

9. Spore preparations with no or very few detached sterigmal remnants; spores mostly finely ornamented 13

10. In alpine zone on calcareous soils, often with *Dryas*; exoperidial spines very delicate .... (*L. frigidum*)

10. In different zones; spines typically less delicate 11

11. Capillitium very thick-walled, to 1.5 µm broad, pores very scattered, spores with coarse, scattered warts; fruitbody turbinate to pyriform, exoperidial spines fairly persistent, pale brown or yellow-brown; very rare or overlooked, in woods, mainly thermophilous oak-woods .... **8. Dark-spored Puffball** (*L. atropurpureum*)

11. Capillitium with thinner walls, less than 1 µm, pores abundant or scattered, spores with dense or scattered warts; fruitbodies either more globose or with more prominent stipe part 12

12. Fruitbody pyriform with cylindrical stipe; spores densely warted; relatively common in neutral to calcareous woodland ..... **9. Soft-spined Puffball** (*L. molle*)

12. Fruitbody subglobose to turbinate; spores mostly densely warted; rare or overlooked, in open, dry grassland .... **10. Steppe Puffball** (*L. decipiens*)

13. Spore deposit yellow-brown; spores weakly ornamented; shiny yellow-brown endoperidium exposed early in development; exoperidium with convergent often dark brown spines; on acid soils, mostly with conifers .... **11. Umber-brown Puffball** (*L. umbrinum*)

13. Spore deposit dark brown; spores conspicuously ornamented; exoperidial spines straight or convergent 14

14. Fruitbody turbinate to pear-shaped with short, variable spines; capillitium with very few septa and tiny pores; very rare or overlooked, with conifers .... **12. Conifer Puffball** (*L. lambinonii*)

14. Fruitbody pestle-shaped or more or less turbinate with straight to convergent fragile spines; capillitum with frequent septa, very fragile; acid soils, open and wooded habitats, very rare .... **13. Heath Puffball** (*L. ericaeum*)

British Puffballs, Earthstars and Stinkhorns

1. STUMP PUFFBALL

Map. No. 52

Lycoperdon pyriforme Schaef.: Pers., *Synops. Meth. Fung.*: 145 (1801); Schaef., *Icon. fung. Bav.* 4: 128 (1774).
Lycoperdon pyriforme var. *excipuliforme* Desmazières, *Crypt. France* ser. I, no. 1152 (1825).
Lycoperdon pyriforme var. *tessellatum* Pers., *Synops. Meth. Fung.*: 148 (1801).
  Additional synonyms: see Kreisel (1962) and Demoulin (1972b).

Selected descriptions: Demoulin (1972b: 187–192); Kreisel (1962: 134–137) .

Selected illustrations: Breitenbach & Kränzlin (1986: 519); Cetto (1988: 337); Dähncke & Dähncke (1979: 573); Marchand (1976: pl. 370); Gerhardt (1985: 198); Jahn (1979: pl. 207); Jeppson (1984: 38, drawing); Lange & Hora (1965: 219); Michael & al. (1986: pl. 149); Moser & Jülich (1989: 1); Wakefield & Dennis (1981: pl. 109, f. 3).

Diagnostic characters: on wood; smooth spores; whitish, firm subgleba; prominent white rhizoids; non-poroid capillitium; spiny spherocysts.

Fruitbodies clustered on decaying wood, pyriform to pestle-shaped, rarely subglobose, 1.5–6 cm tall, connected by thick white rhizoids; opening by fairly large rounded operculum. *Exoperidium* warty-granulose to slightly spiny on stipe, appressed squamulose, soon glabrous apically, exposing papery, matt endoperidium, pale brown to darker reddish brown. *Gleba* with distinct pseudocolumella, white through olive to grey-brown, with strong smell of gas. *Subgleba* firm, areolate, remaining whitish. *Spore deposit* olive-brown.
  Basidiospores almost smooth, 3.5–4 µm diam.; sterigmal remnants absent from mounts. *Capillitium* brown, elastic, non-poroid, walls 0.7–1 µm thick. *Paracapillitium* abundant. *Exoperidium* with large, thick-walled, irregularly shaped, spiny, spherocysts.

Habitat: woodland, parks and gardens, on fairly strongly decayed wood, sometimes on buried wood, mostly hardwood but also softwood; prefers richer, more alkaline soils.

Distribution and frequency: very common throughout the country. Almost cosmopolitan; rare in the Mediterranean region.

Other remarks: more or less globose forms and forms with long stipes and different surface ornamentation have occasionally been separated at variety level but a complete intergrading series seems to exist. Demoulin (1972b) considered *L. pyriforme* to be taxonomically isolated.

Stump Puffball

Fig. 109. *Lycoperdon pyriforme*. (Bedfordshire, Maulden Woods, Nov. 1976, Outen).

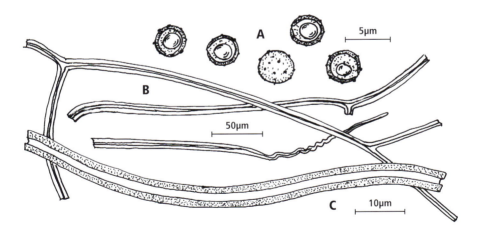

Fig. 110. *Lycoperdon pyriforme* (Derbyshire, Stoke, 3 Oct. 1954, Palmer 342). A, Spores; B, capillitial threads; C, detail of capillitial thread.

145

British Puffballs, Earthstars and Stinkhorns

2. FLAKY PUFFBALL

Map. No. 53

Lycoperdon mammiforme Pers., *Synops. Meth. Fung.*: 146 (1801) [as '*mammaeforme*'].
Lycoperdon velatum Vittad, *Monogr. Lycoperd.*: 43 (1842) (p. 187, in preprint from 1843).
  Misapplications: *Lycoperdon cruciatum* Rostkovius; *Lycoperdon candidum* Pers.: Pers.
  (see Palmer 1968).

Selected descriptions: Demoulin (1972b: 164–165); Kreisel (1962: 156–157).

Selected illustrations: Bolets Catal. 380 (1989); Bon (1987: 305); Bull. Soc. Myc. Fr.
97(3), atlas 226 (1981); Cetto (1988: 343); Dähncke & Dähncke (1979: 571); Jeppson
(1984: 33, drawing); Kobler (1986: opposing p. 30); Marchand (1976: pl. 367); Michael
& al. (1986: 153); Phillips (1981: 247).

Diagnostic characters: flaky veil on top of minutely spinulose exoperidium; often with
pale pinkish colours.

Fruitbodies in small groups or solitary, pyriform, 3–6 (–10) cm tall, without prominent
rhizoids. *Exoperidium* with fugacious, white, veil-like covering, which is shed in more
or less polygonal plates, those at the base persisting longer; another exoperidal layer
underneath consists of a dense, but very fine, sometimes pinkish, spinulose and also
fugacious ornamentation; individual spines converge in small groups. Endoperidium
papery, shiny, olive-brown to more bronzy brown. *Gleba* at maturity umber, with
indistinct pseudocolumella. *Subgleba* well developed, coloured olive-brown or lilaceous
at maturity, alveolate. *Spore deposit* olive-brown to chocolate-brown.
  Basidiospores verrucose, (4–) 4.4–5 (–6.5) µm, intermixed with detached sterigmal
remnants, up to 23 µm long. *Capillitium* elastic, brown, with scattered small pores, walls
1–1.6 µm broad. *Paracapillitium* absent.

Habitat: on humus, in calcareous, deciduous woodland.

Distribution and frequency: rare, mostly western and southern, with a single Irish
record, and not known from Scotland. Central Europe extending into southern
Scandinavia. Replaced by *Lycoperdon floccosum* in North America.

Other remarks: specimens having lost the outer veil resemble *L. molle*. Young, 'veiled'
specimens are unmistakeable.

Flaky Puffball

Fig. 111. *Lycoperdon mammiforme*. (Denmark, 1992, Vesterholt).

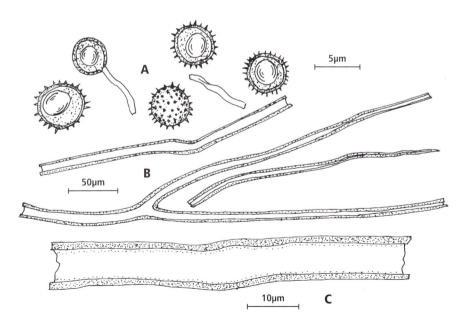

Fig. 112. *Lycoperdon mammiforme*. (Sussex, Arundel, Nov. 1979, Rayner). A, Spores; B, capillitial threads; C, detail of capillitial thread.

3. PEDICELLED PUFFBALL

Map. No. 54

Lycoperdon caudatum Schröt., *Pilze Schlesiens* 1: 698 (1889).
L. pedicellatum Peck in *Bull. Buffalo Soc. Nat. Sci.* 1: 63 (1873), non *L. pedicellatum* Batsch (1783).

Selected descriptions: Dring & Reid (1964: 296–297, as *L. pedicellatum*); Kreisel (1962: 149–151, as *L. pedicellatum*); Demoulin (1972b: 211–213, as *L. pedicellatum*).

Selected illustrations: Moser & Jülich (1989: 7, as *L. pedicellatum*); Ryman & Holmåsen (1984: 589); Phillips (1981: 249, as *L. pedicellatum*); Breitenbach & Kränzlin (1986: 517, as *L. pedicellatum*); Anonymous (1981: frontpage, bottom, as *L. pedicellatum*); Cetto (1988: 341 & 1992: pl.2870, as *L. pedicellatum*); Jeppson (1984: 25, drawings).

Diagnostic characters: long-pedicellate spores; exoperidium with fugaceous, long, composite spines.

Fruitbodies in groups or solitary, turbinate to pyriform, up to 5 cm tall, slightly rooting in debris/humus, without prominent rhizoids. *Exoperidium* with strongly fugacious, cream to yellow-brown, convergent spines, which leave faint, quickly disappearing, areolate pattern; spines often detached in patches. *Endoperidium* pale brown. *Gleba* olive-brown to grey-brown, with indistinct pseudocolumella. *Subgleba* alveolate, lilaceous brown. *Spore deposit* olive-brown to grey-brown.
Basidiospores nearly smooth to asperulate, globose to subglobose, (3.5–) 4–4.5 (–5.2) µm, with attached sterigmal remnant, to 35 µm long. *Capillitium* sub-elastic, yellow to grey-brown, with fairly abundant, rounded pores; walls 0.6–1 µm thick. *Paracapillitium* absent or rare.

Habitat: calcareous dune-slacks and damp calcareous woodland.

Distribution and frequency: rare, only known from a few localities in southern Scotland. A boreal-continental species, also occurring in North America.

Other remarks: this rare puffball is included in several national red data lists and the European red list, where it was placed in the category showing widespread losses and some national extinctions (Ing 1993). It has mostly been listed as *L. pedicellatum*. Based on the literature, the open type of habitat seems to be the most typical for this species. The first British record was published from Dumbartonshire, Scotland by Dring & Reid (1964). The earliest available name is *L. candidum* Pers.: Pers. but since this has been consistently misapplied to another *Lycoperdon* species Demoulin (1970) rejected the name as a '*nomen ambiguum*'. The distribution pattern of *L. norvegicum* Demoulin is very similar and this species should be looked for in suitable Scottish sites.

Pedicelled Puffball

Fig. 113. *Lycoperdon caudatum.* (Scotland, 1991, Laessøe).

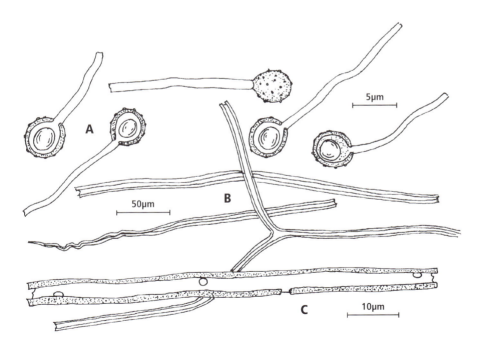

Fig. 114. *Lycoperdon caudatum.* (Dunbarton, Sept. 1963, Taylor). A, Spores; B, capillitial threads; C, detail of capillitial thread.

149

British Puffballs, Earthstars and Stinkhorns

4. HEDGEHOG PUFFBALL

Map. No. 55

Lycoperdon echinatum Pers.: Pers., *Synops. Meth. Fung.*: 147 (1801).
Lycoperdon hoylei Berk. & Broome in *Ann. Mag. Nat. Hist.* ser. 4, 7: 430 (1871).

Selected descriptions: Demoulin (1972b: 183–185); Kreisel (1962: 157–159).

Selected illustrations: Bolets Catal. no. 81; Bon (1987: 305); Cetto (1988: 335); Dähncke & Dähncke (1979: 570); Francis (1990: 42F, young frbs); Jülich (1989: 3); Kobler (1982: opposite 222); Lange & Hora (1965: 219); Marchand (1976: pl. 366); Michael & al. (1986: pl. 152); Moser & Phillips (1981: 246).

Diagnostic characters: very long, brown spines, falling away to leave marked areolate pattern; spore deposit very dark.

Fruitbodies in small groups or solitary, 3–10 cm high, 1–6 cm diam., subglobose, obovoid to pyriform. *Exoperidium* of dense, dark brown, initially white, 3–6 mm long, composite, convergent spines, which fairly easily detach leaving marked areolate pattern on endoperidium. *Endoperidium* papery, pale brown, with polygonal pattern of reddish brown felty mycelium from exoperidium. *Gleba* violaceous grey-brown to chocolate-brown, with indistinct pseudocolumella. *Subgleba* areolate, various shades of brown and lilac. *Spore deposit* chocolate-brown or with lilac tints.
  Basidiospores verrucose, globose, 3.8–5.1 µm. *Capillitium* elastic, pale brown, small rounded pores fairly abundant; wall 0.7–1 µm wide. *Paracapillitium* absent. *Exoperidium* with large, deep brown, irregularly shaped, thick-walled spherocysts.

Habitat: calcareous soil in mainly beech (*Fagus*) woodland.

Distribution and frequency: locally common in SE England and the Cotswolds, very rare elsewhere but recorded as far north as Perthshire in Scotland. Mainly central Europe reaching southern Finland and in the south restricted to moist, higher altitude forests. Replaced by *L. americanum* Demoulin in North America.

Other remarks: often occurs in sites with many species of *Lepiota* (Agaricales) and often with *Lycoperdon mammiforme*. Although very characteristic it has often in the past been confused with *L. nigrescens*.

150

Hedgehog Puffball

Fig. 115. *Lycoperdon echinatum*. (Denmark, Arrhus, 19 June 1993, Lange).

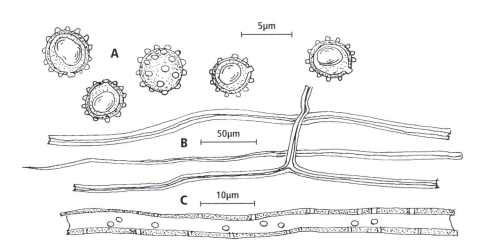

Fig. 116. *Lycoperdon echinatum*. (Gloucestershire, Sept. 1992, Laessøe). A, Spores; B, capillitial threads; C, detail of capillitial thread.

British Puffballs, Earthstars and Stinkhorns

5. COMMON PUFFBALL

Map. No. 56

Lycoperdon perlatum Pers.: Pers., *Synops. Meth. Fung.*: 145 (1801).
Lycoperdon gemmatum var. *perlatum* (Pers.: Pers.) Fr. , *Syst. mycol.* 3: 37 (1829).

Additional synonyms: see Kreisel (1962) and Demoulin (1972b).

Selected descriptions: Calonge & Demoulin (1975: 276–277); Demoulin (1972b: 227–231); Kreisel (1962: 144–146).

Selected illustrations: Bon (1987: 305); Breitenbach & Kränzlin (1986: 518); Clémençon & al. (1980: 284-85); Dähncke & Dähncke (1979: 572); Gerhardt (1985: 194); Ing (1989: 127B); Moser & Jülich (1989: 8); Phillips (1981: 248); Ryman & Holmåsen (1984: 591); Wakefield & Dennis (1981: pl. 110, f. 2).

Diagnostic characters: reticulate pattern on endoperidium; fugacious, conical, non-convergent exoperidial spines surrounded by ring of warts; spores typically less than 4 µm in diam.

Fruitbodies solitary or in dense clusters, subglobose, pyriform, sub-cylindric or almost pestle-shaped, 2–9 cm tall, 2–4 cm diam., white, cream to pale brown; with branched, thin rhizoids which hold the substratum. *Exoperidium* consisting of 1–2 mm long, fragile, white, cream to pale brown, conical spines, surrounded by a persistent, circular row of warts; spine scars and warts produce characteristic reticulate pattern on endoperidium; pattern much less evident or absent on pseudo-stipe. *Endoperidium* papery, grey-brown. *Gleba* white then brown to olive-brown with well-developed pseudocolumella. *Subgleba* strongly developed, alveolate, olive-brown to grey brown. *Spore deposit* yellow-brown, olive-brown to grey-brown.
  Basidiospores verrucose, (2.9–) 3.1–4.1 (–4.4) µm, sometimes with some detached sterigmal remnants in mounts. *Capillitium* elastic, yellow-brown, 3–7.5 µm wide, walls comparatively thin; pores abundant to rare, mainly present in peripheral capillitium, of variable size. *Paracapillitium* mostly abundant. *Exoperidium* with spherocysts, 20–30 µm, thin-walled (< 1 µm).

Habitat: occurs in deciduous and coniferous woods and plantations. It has a slight preference for richer soils. Occasionally, it occurs on dead wood but confusion with *L. pyriforme* should not be possible.

Distribution and frequency: uniformly distributed over the British Isles and very common. Distributed over the greater part of the world (sub-cosmopolitan).

Other remarks: the equally common *L. nigrescens* has often been referred to as a variety of *L. perlatum*. The former is distinguished mainly by its more persistent and much darker spines. Furthermore, the spores are larger and less ornamented. *Lycoperdon norvegicum* Demoulin is a boreal-continental species, which differs from *L. perlatum* in having small, almost smooth spores and also in having more delicate, gracile spines (Demoulin 1971; Jeppson 1984). It should be looked for in Scotland.

Common Puffball

Fig. 117. *Lycoperdon perlatum*. A, Young fruitbodies (Surrey, Virginia Water, 7 Oct. 1993, Legon); B, old fruitbodies (Devon, Bovey Tracey, 5 Oct. 1985, Roberts).

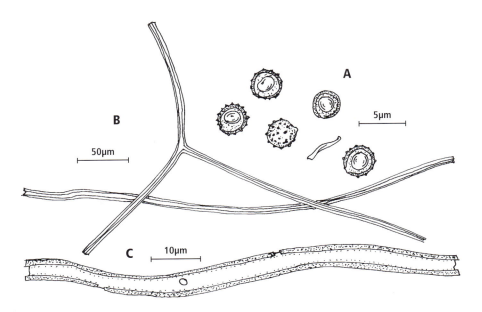

Fig. 118. *Lycoperdon perlatum*. (Buckinghamshire, May 1963, Reid). A, Spores; B, capillitial threads; C, detail of capillitial thread.

6. BLACKISH PUFFBALL

Map. No. 57

Lycoperdon nigrescens Pers., *Neues Mag. Bot.* 1: (1794).
Lycoperdon perlatum var. *nigrescens* (Pers.: Pers.) Pers., *Synops. Meth. Fung.*: 146 (1801).
Lycoperdon foetidum Bonord., *Handb. allgem. Mykol.*: 253 (1851).
   Misapplication: *L. hoylei* Berk. & Broome (Palmer 1968).

Selected descriptions: Calonge & Demoulin (1975: 273–274); Demoulin (1972b: 222–224, as *L. foetidum*); Kreisel (1962: 147–149, as *L. foetidum*).

Selected illustrations: Cetto (1983: pl.1191); Gerhardt (1985: 195); Phillips (1981: 248, as *L. foetidum*); Wakefield & Dennis (1981: pl. 110, f. 3).

Diagnostic characters: dark, convergent exoperidial spines shorter than 3 mm, leaving a reticulate pattern; spores mostly larger than 4 µm diam.

Fruitbodies in small groups, subglobose to pyriform, 2–5 cm high, 1.5–4 cm diam., cream to blackish brown with paler endoperidium visible underneath. *Exoperidium* with sub-persistent, mostly converging, pale brown, brown to blackish brown, 1–3 mm long, cylindrical to slightly conical spines, surrounded by a felty layer, leaving a reticulate pattern on overmature specimens. *Endoperidium* cream to pale brown, matt, papery. *Gleba* olive-brown or other shades of brown, with well developed pseudocolumella; smell of gas or 'metal' (as in *L. pyriforme*). *Subgleba* coarsely alveolate, shades of brown, occasionally tinted olive or lilac. *Spore deposit* yellow-brown, olive-brown to brown.
   Basidiospores verrucose, globose to subglobose, (3.4–) 3.8–4.4 (–4.8) µm. *Capillitium* elastic, yellow-brown with many pores, especially in periphery of gleba, wall 0.7–0.9 µm thick. *Paracapillitium* mostly present, especially in periphery of gleba. *Exoperidium* with simple spherocysts.

Habitat: on very humose sand; acidic woodland, heathland and grassland including moss-rich lawns on acid soils.

Distribution and frequency: throughout the British Isles; common. All over Europe but with suboceanic pattern and with similar trend in North America.

Other remarks: the change in starting point for botanical nomenclature has made it necessary to revert to the much older name *L. nigrescens* instead of the name made more familiar by modern treatments, *L. foetidum*. The ecological requirements of this species are similar to those of *L. umbrinum* but *L. nigrescens* is less attached to conifers. *Lycoperdon perlatum* tends to be more base-loving, occuring in richer sites.

Blackish Puffball

Fig. 119. *Lycoperdon nigrescens.* (1992, Laessøe).

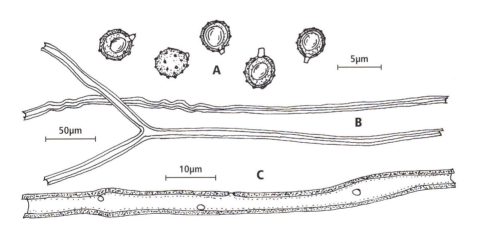

Fig. 120. *Lycoperdon nigrescens.* (Sussex, Coldwaltham, Sept. 1971, Reid). A, Spores; B, capillitial threads; C, detail of capillitial thread.

7. GRASSLAND PUFFBALL

Map. No. 58

Lycoperdon lividum Pers. in *J. Bot. Paris* 2: 18 (1809).
L. spadiceum Pers., in *J. Bot. Paris* 2: 20 (1809) non *L. spadiceum* (Schaeff.: Pers.) Poiret 1808.

Selected descriptions: Calonge & Demoulin (1975: 274); Demoulin (1972b: 193–195); Kreisel (1962: 137, as *L. spadiceum*); Monthoux & Röllin (1984: 193–195); Ortega & al. (185: 142–143, incl. SEM of spores).

Selected illustrations: Cetto (1992: pl. 2868); Phillips (1981: 249, as *L. spadiceum*); Ryman & Holmåsen (1984: 592); Wakefield & Dennis (1981: pl. 109, f.2, as *L. spadiceum*).

Diagnostic characters: exoperidium granulose without distinct spines; spores almost smooth; fragile capillitium with big pores; in open, mostly calcareous grasslands.

Fruitbodies in small groups, more or less capitate with well delimited pseudostipe, which adheres to ball of substratum, 1–4 cm tall, 1–2.5 cm diam., yellow-brown to relatively pale grey-brown. *Exoperidium* of pale brown granules or slightly spinulose on pseudostipe, fugacious and hardly visible in well-matured specimens. *Endoperidium* papery, pale grey-brown. *Gleba* in shades of olive-yellow and olive-brown; pseudocolumella indistinct. *Subgleba* alveolate, spongy, brown or with lilac tints. *Spore deposit* olive, olive-brown or yellow-brown.

Basidiospores asperulate, (3.3–) 3.7–4.4 (–4.9) μm, globose to subglobose, with tiny sterigmal remnant attached, no detached remnants in mounts. *Capillitium* fragile, yellow-brown, 4–8 μm wide, pores abundant, of various shapes. *Paracapillitium* absent or rarely present in limited amount. *Exoperidium* with simple spherocysts.

Habitat: on weakly acid to calcareous humus or soil in dry grassland sites, especially coastal grassland and short turf.

Distribution and frequency: widespread in the British Isles; common, especially in coastal areas. Common in similar situations on the European continent and with a distinct distribution pattern in North America.

Other remarks: often overlooked, seemingly due to the habitat being less visited by recorders or because this species has very nondescript fruitbodies making field identification difficult. The illustration in Breitenbach & Kränzlin (1986) could according to Demoulin (1987) be *Bovista aestivalis*.

Grassland Puffball

Fig. 121. *Lycoperdon lividum.* (Middlesex, Twickenham, Aug. 1992, Brown).

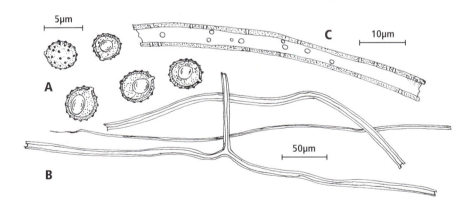

Fig. 122. *Lycoperdon lividum.* (Middlesex, Twickenham, Aug. 1992, Brown). A, Spores; B, capillitial threads; C, detail of capillitial thread.

British Puffballs, Earthstars and Stinkhorns

8. DARK-SPORED PUFFBALL

Map No. 59

Lycoperdon atropurpureum Vittad., *Monogr. Lycoperd.*: 42 (1842) (p. 186 in preprint, 1843).

Selected descriptions: Calonge & Demoulin (1975: 270); Demoulin (1972b: 159–160); Jeppson & Demoulin (1989: 131–134, includes photographs).

Selected illustrations: Moser & Jülich (1989: 2); Jeppson (1987: 276, drawing).

Diagnostic characters: spores coarsely warted; elastic thick-walled capillitium (to 1.5 µm) with few small pores; endoperidium barely visible under well-formed, erect, brown, exoperidial spines. Compare with *L. decipiens* and *L. molle*.

Fruitbodies in small groups or solitary, subglobose, turbinate to pyriform, 2–5 cm high, binding substrate with rhizoids and mycelium; yellow-brown. *Exoperidium* typically with well-formed, slender, erect, fragile, connected or simple, brown spines. *Endoperidium* not conspicuous, only exposed in patches, shiny, cream-coloured. *Gleba* with indistinct pseudocolumella. *Subgleba* alveolate with fairly small locules, to 0.8 mm diam., slowly colouring from yellow to chocolate-brown. *Spore deposit* chocolate-brown with purplish tinges.

  Basidiospores coarsely verrucose, verrucae relatively well-spaced and almost truncate, (4.1–) 4.4–5.5 (–5.8) µm; detached sterigmal remnants normally abundant in mounts. *Capillitium* elastic, reddish brown, 4.5–7 µm diam., walls thick, 1–1.5 µm, with few, small pores in centre of gleba, more in top of pseudocolumella. *Paracapillitium* absent. *Exoperidium* with simple spherocysts, thick-walled at base of spines, thin-walled apically.

Habitat: thermophilic woods, like southern oak woodland.

Distribution and frequency: southern; very rare but could have been overlooked due to confusion with *L. molle*. Common in Mediterranean and sub-Mediterranean oak woods, with few isolated occurrences as far north as southern central Sweden.

Other remarks: *Lycoperdon mauryanum* Demoulin is apparently a North American equivalent to *L. atropurpureum* (Demoulin 1972). Kreisel (1962) and Palmer (1968) included *L. atropurpureum* under *L. molle* but Kreisel (1973) accepted the species with *L. decipiens* as a synonym. Literature references to this species, especially older ones, should not be accepted at face value. Most records from Britain were redetermined by Demoulin (1972b).

Dark-spored Puffball

Fig. 123. *Lycoperdon atropurpureum.* A, (Austria, Almersberg, 25 Aug. 1987, Mrazek); B, section (Austria, Almersberg, 3 Sept. 1982, Mrazek).

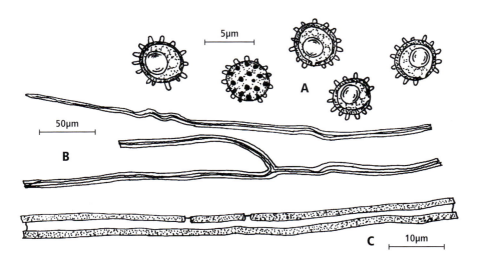

Fig. 124. *Lycoperdon atropurpureum.* (Italy, Vittadini, K, lectotype). A, Spores; B, capillitial threads; C, detail of capillitial thread.

9. SOFT-SPINED PUFFBALL

Map No. 60

Lycoperdon molle Pers.: Pers., *Synops. Meth. Fung.*: 150 (1801).

Selected descriptions: Calonge & Demoulin (1975: 275–276); Demoulin (1972b: 168–173).

Selected illustrations: Breitenbach & Kränzlin (1986: 516); Cetto (1992: 2866); Gerhardt (1985: 1975); Jeppson (1984: 34); Marchand (1976: 368).

Diagnostic characters: fruitbody pyriform with pronounced pseudostipe; typically café-au-lait with delicate spines; spores markedly and densely verrucose, intermixed with sterigmal remnants; capillitium with many small pores.

Fruitbodies in small groups or solitary, turbinate to stipitate pyriform, 2–5 (–9) cm high, 1–3.5 (–7) cm diam., binding substratum with thin rhizoids and mycelium, cream to café-au-lait. *Exoperidium* with variable ornamentation, typically with thin, grey-brown, straight, short spines intermixed with granules and longer, up to 1.5 mm, often convergent, fragile spines, especially at fruitbody base; occasionally totally devoid of spines. *Endoperidium* mostly covered, papery, cream or pale yellow-brown. *Gleba* olive-brown. *Pseudocolumella* present, but not distinct. *Subgleba* prominent, alveolate, locules 0.5–1 mm diam., brown with lilac tints. *Spore deposit* chocolate-brown.

Basidiospores verrucose, (3.8–) 4–5.1 (–6.2) μm, intermixed with sterigmal remnants. *Capillitium* elastic, in shades of brown, mostly 4–7 μm diam. but some threads up to 10 μm diam., wall 0.7–1.4 μm thick; pores abundant, typically small and rounded but very variable. *Paracapillitium* mostly absent. *Exoperidium* with simple spherocysts.

Habitat: on neutral to calcareous soils, mostly in woodland but can also be found in open situations.

Distribution and frequency: mostly southern but widespread; fairly common. Widespread in Europe, occurring at least as far north as Iceland, and North America, especially in temperate regions.

Other remarks: the habitat range is very broad and compares with that of *L. perlatum*. *Lycoperdon frigidum* Demoulin is a closely related species from arctic-alpine localities, which possibly could occur in Scotland or in the Welsh mountains. It is often associated with *Dryas* and differs principally from *L. molle* in the larger, less ornamented spores and in some capillitial features. *Lycoperdon niveum* Kreisel is also arctic and could occur at high altitude in the British Isles (Demoulin 1972b). Only recently has the name *L. molle* been applied in British literature. Earlier workers mainly either misapplied *L. atropurpureum* or *L. umbrinum* to *L. molle*.

Soft-spined Puffball

Fig. 125. *Lycoperdon molle*. (Oct. 1993, Laessøe).

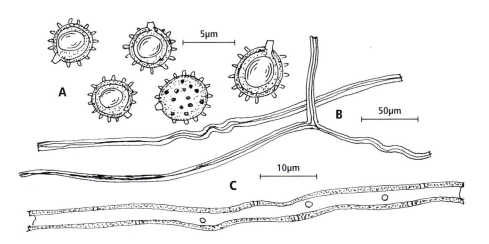

Fig. 126. *Lycoperdon molle*. (Gloucestershire, Sept. 1992, Laessøe). A, Spores; B, capillitial threads; C, detail of capillitial thread.

10. STEPPE PUFFBALL

Map No. 61

Lycoperdon decipiens Dur. & Mont., *Expl. Sci. Algérie,* Bot. 1. Crypt.: 380 (1848) .

Selected descriptions: Calonge & Demoulin (1975: 271–272); Demoulin (1972b: 156–158); Jeppson & Demoulin (1989: 134).

Selected illustrations: Jeppson (1987: 277, drawing); Moser & Jülich (1989: 4).

Diagnostic characters: similar to *L. molle* and *L. atropurpureum*; apically often with star-like scales; spores strongly, and typically densely, verrucose; fairly fragile capillitium with abundant, large, irregularly-shaped pores; subgleba poorly developed.

Fruitbodies solitary or in small groups, subglobose turbinate, 1.5–4.5 cm diam., normally wider than high, binds substrate with mycelium and normally a thick rhizoid; cream. *Exoperidium* with fragile, convergent groups of yellow-white to pale brownish spines; apically often appresed star-like. *Endoperidium* cream to grey-brown, matt, visible through exoperidium. *Gleba* with poorly differentiated pseudocolumella. *Subgleba* rather poorly developed, alveolate, cream to chocolate-brown. *Spore deposit* chocolate-brown.

   Basidiospores strongly verrucose, mostly densely so, (4.2–) 4.7–5.6 (–6) μm, mixed with detached sterigmal remnants. *Capillitium* fragile to subelastic, brown, walls relatively thin, 0.6–1 μm; pores usually abundant and irregular in shape; centre of gleba with many isolated hyphae (*Bovista*-type). *Paracapillitium* absent. *Exoperidium* of simple spherocysts, with no difference between base and apex of spines.

Habitat: on turf in dry, mostly calcareous, grassland; thermophilous.

Distribution and frequency: rare in Britain, known from a few localities in the south and west. Widespread in Europe reaching further north than *L. atropurpureum*.

Other remarks: see remarks under *L. atropurpureum* and *L. molle*.

Steppe Puffball

Fig. 127. *Lycoperdon decipiens*. (Sweden, Skållerud, 25 Sept. 1988, Jeppson).

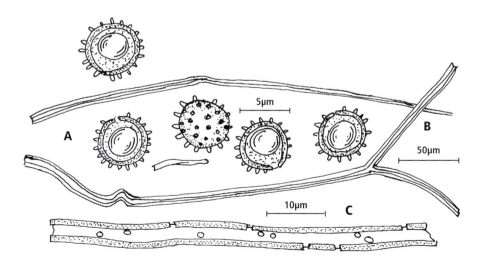

Fig. 128. *Lycoperdon decipiens*. (Sussex, Goodwood, 23 Aug. 1980, Reid). A, Spores; B, capillitial threads; C, detail of capillitial thread.

11. UMBER-BROWN PUFFBALL

Map No. 62

Lycoperdon umbrinum Pers.: Pers., *Synops.Meth. Fung.*: 147 (1801).

Selected descriptions: Calonge & Demoulin (1975: 277–278); Demoulin (1972b: 205–208); Kreisel (1962: 139–141); Ortega & al. (1985: 142–143, incl. SEM of spores).

Selected illustrations: Breitenbach & Kränzlin (1986: 520); Marchand (1976: pl. 371); Dähncke & Dähncke (1979: 574); Ryman & Holmåsen (1984: 591); Michael & al. (1986: pl. 151); Gerhardt (1985: 196); Jeppson (1984: 39, drawing).

Diagnostic characters: exoperidial spines brown, of equal length, persistent; endoperidium shiny, yellowish; spore deposit yellow-brown; spores asperulate, not mixed with sterigmal remnants; capillitial pores large; mostly with conifers.

Fruitbodies in small groups or solitary, shape variable, mostly pyriform, 2–5 cm high, 1–4 cm diam., pale brown to reddish brown. *Exoperidium* densely clad in uniform, pale brown, reddish brown to blackish brown, slender, convergent spines, to 1 mm long, fairly persistent but shed apically; not mixed with granules. *Endoperidium* visible between exoperidial ornamentation and clearly so apically, when spines are shed;, cream to yellowish. *Gleba* olive-brownish, with fairly distinct pseudocolumella. *Subgleba* alveolate, olive, grey-brown or grey-lilac. *Spore deposit* yellow-brown.

Basidiospores asperulate, (3.7–) 4.3–5.1 (–5.6) μm with short attached sterigmal remnant; no sterigmal remnants in mounts or very rarely with some. *Capillitium* elastic, yellow-brown, 4.4–10 μm diam., walls thin to medium-thick; pores mostly abundant and large. *Paracapillitium* scarce to abundant.

Habitat: on acid soils, mainly under conifers, e.g. in spruce (*Picea*) plantations.

Distribution and frequency: widespread but rare in Britain. Fairly common on the European continent; suboceanic distribution, lacking in continental parts.

Other remarks: the spore ornamentation is finer than that of the similar *L. atropurpureum*, *L. decipiens* and *L. molle*. *Lycoperdon estonicum* Demoulin is apparently also very similar, differing in having spores with attached, long sterigmal remnants. It has not been recorded in the British Isles. Old records of *L. umbrinum* often represent *L. molle*.

Umber-brown Puffball

Fig. 129. *Lycoperdon umbrinum*. (Sweden, Västmanland, 20 Aug. 1988, Jeppson).

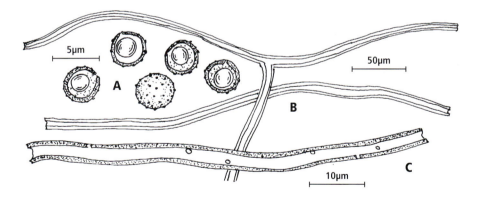

Fig. 130. *Lycoperdon umbrinum*. (Germany, Oberfranken, 25 Aug. 1960, Palmer). A, Spores; B, capillitial threads; C, detail of capillitial thread.

12. CONIFER PUFFBALL

Map No. 63

Lycoperdon lambinonii Demoulin in *Lejeunia* 62: 13 (1972).

Selected descriptions: Demoulin (1972a: 13; 1972b: 173–178); Martín Esteban (1988: 309–311); Ortega & al. (1985: 144–145, incl. of SEM of spores).

Selected illustrations: Jeppson (1984: 31, drawing).

Diagnostic characters: very similar to *L. molle* and *L. umbrinum*; small spores; few, tiny pores in capillitium.

Fruitbodies in groups or solitary, subglobose, turbinate or pyriform, 2–5 cm high, 1–3.5 cm diam., brown or yellowish brown, binding substratum with hyphae. *Exoperidium* with variable ornamentation, granulose and/or with thin, short (less than 0.8 mm), fragile, sometimes convergent spines, cream to yellowish brown or darker brown. *Endoperidium* hardly visible. *Gleba* with well developed pseudocolumella. *Subgleba* alveolate with locules 0.3–0. 7 mm, pale brown, with or without lilac tinges. *Spore deposit* brown.

Basidiospores asperulate, (3.4–) 3.8–4.6 (–4.8) µm, with detached sterigmal remnants in mounts. *Capillitium* elastic, brown, 4–7 µm diam., pores mostly rare and tiny. *Paracapillitium* absent, occasionally present. *Exoperidium* with simple spherocystes.

Habitat: on humus under conifers and in more open situations; possibly with preference for calcareous soils.

Distribution and frequency: very rare in the British Isles, known only from a single Irish record, but could easily have been overlooked. Widespread in Europe and North America.

Other remarks: it was predicted by Demoulin & Marriott (1981) that this very widespread species would occur in the British Isles. Demoulin (1972a,b) compared the species with *L. molle* and *L. umbrinum* stating that it occupies an intermediate position. Jeppson (1984) found it very close to *L. molle*.

Conifer Puffball

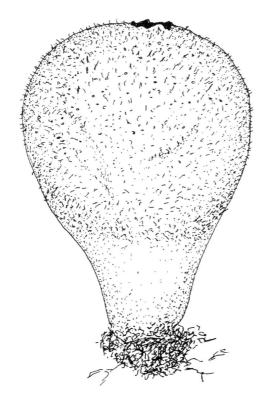

Fig. 131. *Lycoperdon lambinonii.* (after Jeppson, 1984).

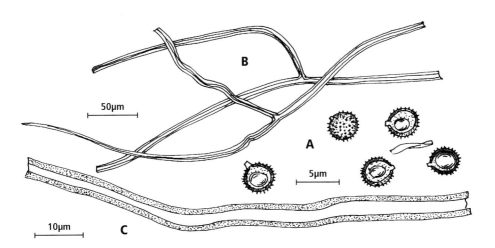

Fig. 132. *Lycoperdon lambinonii.* (Tenerife, La Laguna, 12 Jan. 1973, Beltran Tejera). A, Spores; B, capillitial threads; C, detail of capillitial thread.

13. HEATH PUFFBALL

Map No. 64

Lycoperdon ericaeum Bonord. in *Bot. Zeit.* 15: 596 (1857).
Lycoperdon muscorum Morgan in *Journ. Cincinnati Soc. Nat. Hist.* 14: 16 (1891).
  Additional synonyms: see Demoulin (1972b).

Selected descriptions: Calonge & Demoulin (1975: 273); Demoulin (1972b: 200–204); Kreisel (1962: 141–144, as *L. muscorum*).

Selected illustrations: Cetto (1992: pl. 2867); Moser & Jülich (1989: 5).

Diagnostic characters: fragile capillitium with very evident pores; no sterigmal remnants in mounts; exoperidium with pale, poorly developed, fragile spines.

Fruitbodies solitary or in small groups, mostly pyriform (var. *subareolatum* with long pseudostipe), cream to café-au-lait. *Exoperidium* with poorly developed, fragile, convergent, often appressed and stellate spines, often with pulverulent granules intermixed. *Endoperidium* not very exposed, cream. *Gleba* with poorly developed pseudocolumella. *Subgleba* alveolate, typically lilaceous brown. *Spore deposit* brown (paler than chocolate-brown).
  Basidiospores asperulate, (3.8–) 4–4.7 (–5.3) µm, without attached or detached sterigmal remnants. *Capillitium* fragile, yellow-brown, 4.5–8 µm diam., wall 0.5–0.8 µm thick; pores abundant, large and normally regular in outline. *Paracapillitium* absent. *Exoperidium* with simple spherocysts.

Habitat: on acid, humid turf in open habitats. In North America occurring in a wider habitat range.

Distribution and frequency: very rare or overlooked. Subcontinental distribution in Europe, also occurs in North America.

Other remarks: *Lycoperdon ericaeum* var. *subareolatum* (Kreisel) Demoulin has been recorded over most of the European continent and in North America but it has apparently not been recorded in Britain. The varieties mainly differ in shape of the fruitbodies and in habitat requirements. *Lycoperdon muscorum* is apparently a synonym of *L. ericaeum var. subareolatum*, but if the varieties were separated at the species level the name *L. muscorum* would be the name to use.

Heath Puffball

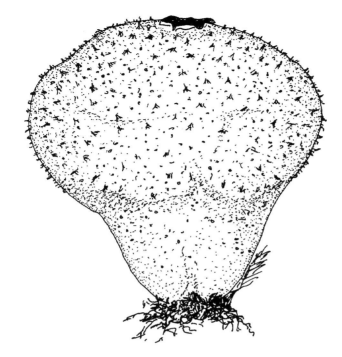

Fig. 133. *Lycoperdon ericaeum.* (after Jeppson, 1984).

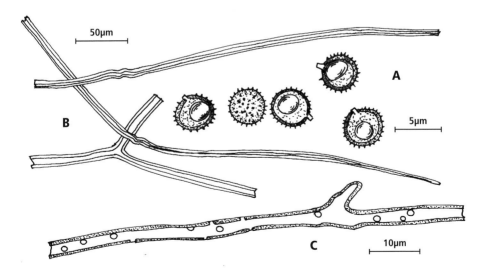

Fig. 134. *Lycoperdon ericaeum.* (Germany, Brandenburg, Triglitz). A, Spores; B, capillitial threads; C, detail of capillitial thread.

STINKHORNS AND CAGE FUNGI

PHALLALES G. Cunn.
in *Proc. Linn. Soc. N.S.W.* 56: 3 (1931).

Fruitbody comprising a gleba, borne on a receptacle, initially enclosed by a peridium, and remaining so in *Hysterangiaceae*; at first hypogeous, globose to ovoid, either indehiscent or epigeous and dehiscent; with conspicuous rhizomorphs. *Peridium* smooth, membranous, pale, often loose and detersile. *Gleba* embedded in a gelatinous matrix, attached to the peridium either basally or by a series of peridial sutures; pale ochraceous to olive-green or brown, deliquescent, autolysing into a foetid mucilage. *Receptacle* present in epigeous species, subtending the gleba, pseudoparenchymatous, spongy and rapidly expanding on release from within the peridium, ranging from cylindrical and resembling a stipe, to stellate or reticulate.

Hyphal system monomitic, often gelatinized; generative hyphae with clamp-connexions. *Basidiospores* elongate ellipsoid to oblong-cylindrical, hyaline to pale greenish brown, smooth, with a thin to slightly thickened wall, basally truncated by a broad hilar attachment. *Basidia* small, tetra- to octosporic. *Hymenophoral trama* regular, hyaline, gelatinized. *Peridiopellis* an epicutis, at times pseudoparenchymatous, usually subtended by a gelatinous mesoperidium. *Development* angiocarpic. *Type family*: *Phallaceae* Corda.

The *Phallales* fruitbody opens in epigeous species to release a rapidly expanding sterile structure, called the receptacle, which bears the spore-producing gleba above ground. The gleba autolyses to a mucilaginous slime and develops an extremely unpleasant, foetid odour in order to attract flying insects which enable spore dispersal. The *Hysterangiaceae* E. Fisch., or Stinkhorn Truffles, contain the hypogeous genus, *Hysterangium* Vittad., which is not considered further here, but a full account may be found in Pegler, Spooner & Young (1993).

Key to the British Families

1. Fruitbody unipileate, with the gleba borne externally on upper part of unbranched, cylindrical, hollow receptacle .... **Stinkhorns** (*Phallaceae*)
1. Fruitbody multipileate, with the gleba borne on inner surfaces of a reticulate to stellate receptacle .... **Cage Fungi** (*Clathraceae*)

PHALLACEAE Corda,
Icon. Fung. 5: 29 (1842, ut 'Phalloideae').

Immature fruitbody ('myco-egg') ovoid, comprising a smooth, pale peridium enclosing a gelatinized gleba and unexpanded receptacle. *Mature fruitbody* unipileate, with a stipe-like receptacle, supporting an apical, mucilaginous gleba. *Receptacle ('stipe')* unbranched, cylindrical, hollow, with spongy texture which rapidly expands at maturity. *Indusium* mostly absent, sometimes present and netlike. *Type genus*: *Phallus* Pers.

Once the peridium has ruptured, growth and elongation of the receptacle is extremely rapid, reaching maximum size within a few hours, or two days at the most. Expansion is effected by utilizing the water store in the gelatinous layer of the peridium. Fruitbodies are quickly located by the nauseous odour of the mucilaginous gleba, which may be detected from a considerable distance. The odour attracts flying insects to the gleba, thereby ensuring spore dispersal. In older fruitbodies, the gleba is completely removed revealing the pale-coloured, costate pileus.

Key to the British Genera

1. Receptacle with a separable, campanulate, apical pileus, with surface ridges, supporting the gleba .... **Stinkhorns** (*Phallus*)
1. Pileus absent and gleba borne subapically over surface of tapering receptacle .... **Dog Stinkhorns** (*Mutinus*)

STINKHORNS

PHALLUS Hadr. Jun.: Pers.,
Synops. Meth. Fung. : 242 (1801, *nom. cons.*); Hadrianus Junius, *Phalli fung. gen. Holland.* (1562).

Dictyophora Desv. in *J. Bot., Paris* 2: 92 (1809).
Hymenophallus Nees, *Syst. Pilze Schwämme*: 251 (1817).
Ithyphallus (Fr.) E. Fisch. in *Journ. Bot. Gard., Berlin* 4 : 41(1886).

Immature fruitbody hypogeous, globose to ovoid, soft-gelatinous, with conspicuous white rhizomorphs. *Peridium* white to pale, smooth, firm-membranous. *Gleba* attached to outer surface of pileus, dark olivaceous to blackish brown. *Receptacle* stipitiform, cylindrical, hollow, pseudoparenchymatous, bearing an apical, campanulate pileus with irregularly branching ridges over the outer surface. *Indusium* absent or scarcely developed in British species. *Spore mass* pale greenish yellow. *Basidiospores* small, ellipsoid, smooth, subhyaline. *Habitat*: amongst leaf litter in damp woodland; rhizomorphs attached to buried wood. *Type species*: *Phallus impudicus* L.: Pers.

An indusiate structure is present on most fruitbodies, although mostly this is reduced to a minute, rudimentary development which is completely hidden by the campanulate pileus. *Phallus* is a cosmopolitan genus, although the species diversity is much greater in tropical areas.

Key to the British Species

1. Peridium white when buried, soon discolouring pinkish mauve on exposure; pileal disk with a crenulated margin; confined to sand-dunes .... **Dune Stinkhorn** (*P. hadriani*)
1. Peridium white, not noticeably bruising pinkish mauve on exposure; pileal disk with a non-crenulated margin .... **Common Stinkhorn** (*P. impudicus*), 2

2. Indusium absent or very reduced below the pileus .... var. ***impudicus***
2. Indusium present and well developed, latticed, with mesh of uniform size .... var. ***togatus***

COMMON STINKHORN

Map No. 65

Phallus impudicus *L.: Pers.*, *Synops. Meth. Fung.* : 242 (1801); Linn., *Spec. Plant.*: 1648 (1753).
Phallus foetidus Sow., *Col. Fig. Engl. Fung.* 3: pl. 329 (1803).
Ithyphallus impudicus (L.: Pers.) E. Fisch. in *Jahrb. K. Bot. Gard., Berlin* 4 : 43 (1886).

Selected descriptions: Berkeley (1860: 297 298, pl.20/3); Bolton (1788: pl.92); Massee (1889: 88, fig.44); Ramsbottom (1953: 179–182); Rea (1922: 23).

Selected illustrations: Breitenbach & Kränzlin (1986: pl. 528); Gerhardt (1985: 215); Phillips (1981: 256); Sarasini (1992: pl.1, 9); Wakefield & Dennis (1981: 202, pl.108/1)

Diagnostic characters: white, unbruising peridium; cylindrical, spongy receptacle; campanulate, ridged pileus; pileal disk not crenulate; indusium absent or reduced.

Unexpanded fruitbody 3–4 (–6) cm diam., globose to ovoid, subepigeous ('witches egg'); exoperidium white to pale cream, smooth, membranous with a gelatinous central layer; splitting apically into 4–5 lobes, retained as a cupulate volva at receptacle base, attached by a stout, white, mycelial cord, 1–1.5 mm diam. *Receptacle: stipe* 15–20 cm tall, 1.5–3 cm diam., cylindrical but tapering below, white, spongy, fragile, hollow; *pileus* glebiferous, attached to apical expansion of stipe, 3–5 cm diam., campanulate, externally pale grey to brownish, reticulate-costate, forming large, angular chambers, 1–3 mm deep; apical disk truncated, usually perforate, whitish, with a non-crenulated margin. *Indusium* absent or slight, attached between stipe and pileal attachment. *Gleba* soon mucilaginous, translucent, greenish black to dark olive-green, with strong foetid odour at maturity. *Spore mass* pale olive-brown.
Basidiospores 4–5.6 x 1.8–2.8 µm.

Habitat: mainly in deciduous woodland, also in conifer plantations, dunes, gardens, in damp situations.

Distribution and frequency: common throughout Britain, occurring singly or in groups from June to October. North temperate.

var. **togatus** (Kalchbr.) Cost. & Dufour, *Nouv. Fl. Champ.*: 288 (1895).

Map No. 66

Hymenophallus togatus Kalchbr. in *Ungar. Akad. Wissensch., Budapest.* 13 (8): 6 (1884).
Phallus impudicus var. *pseudoduplicatus* O. Anders. in *Svensk Bot. Tidskr.* 83: 233 (1989).
Misapplied name: *Dictyophora duplicata* sensu auct.

Selected illustrations: Michael, Hennig & Kreisel (1986: pl.139); Ramsbottom (1953: pl. 10).

Differs from the typical variety by a well developed indusium, up to 4–5 cm long.

Habitat: similar to var. *impudicus*.

Distribution and frequency: recorded many times, although never frequent, from Britain and northern Europe (Handke 1963), mostly incorrectly as *Dictyophora duplicata* (Bosc) E. Fischer (= *Phallus duplicatus* Bosc).

Common Stinkhorn

Other remarks: Andersson (1989) defined the two indusiate taxa when he proposed the variety *pseudoduplicatus*, clearly a later synonym of var. *togatus*. *Phallus duplicatus* is generally more robust, the exoperidium varies from white, yellow to flesh-colour, sometimes with a brownish tinge, and the indusium is larger with a mesh that becomes smaller towards the distal, membranous border. In *P. impudicus* var. *togatus*, the exoperidium remains constantly white, and the shorter indusium has a mesh of uniform size and lacks a membranous border. The white exoperidium relates the variety to *P. impudicus* rather than to *P. duplicatus*, the latter confined to North America and Japan. Rea (1922) listed the variety, under its correct name, commenting that it was rare. Ramsbottom (1953) cited specimens, as *D. duplicata*, from Yorkshire collected in 1915, with an illustration (pl.10) from the New Forest.

Fig. 135. *Phallus impudicus* (Hertfordshire, Prae Wood, Oct. 1989, Outen). **Fig. 136.** *Phallus impudicus*, myco-egg stage (Surrey, Spooner). **Fig. 137.** *Phallus impudicus* (Hertfordshire, Watford, Herb. Plowright). Spores. **Fig. 138.** *Phallus impudicus* var. *togatus* (Denmark, Rønshoved, Sept. 1991, Dickson).

DUNE STINKHORN

Map No. 67

Phallus hadriani *Vent.: Pers.*, *Synops. Meth. Fung.*: 246 (1801); Ventenat in *Mém. Inst. Nat. Sci. Arts* 1: 517 (1798).
Phallus iosmus Berk. apud Smith, *Engl. Fl. Fung.*: 227 (1836).
Kirchbaumia imperialis Schulz. in *Ver. K. Wien Zool.-Bot. Ges.*: 798 (1866).
Ithyphallus impudicus var. *iosmus* (Berk.) E. Fisch. in *Jahrb. K. Bot. Gard., Berlin* 4: 44 (1886).

Selected descriptions: Ramsbottom (1953: 183–184); Rea (1922: 24 as *P. impudicus* var. *iosmus* and *P. imperialis*)

Selected illustrations: Bon (1987: 301); Breitenbach & Kränzlin (1986: pl.527); Elborne (1989: 3, 9); Sarasini (1992: pl.2, 8).

Diagnostic characters: peridium discolouring pinkish mauve; cylindrical, spongy receptacle; pileal disk with crenulated edge; confined to coastal dunes.

Unexpanded fruitbody 3–7 cm diam., globose to ovoid, white when buried in sand but rapidly discolouring flesh-pink to reddish on exposure, attached by a pinkish to violaceous pink mycelial cord. *Receptacle: stipe* 10–25 cm long, 2–3 cm diam., cylindrical, white to very pale greyish, often pinkish to pale red towards the base, spongy, fragile, hollow; *pileus* 3–5 cm diam., campanulate, with reticulate-ridged surface, disk with a crenulate to dentate margin. *Gleba* when fresh having a slight odour, said to resemble that of violets, finally foetid.

Basidiospores 3.2–6 x 2–3.2 µm, elongate ellipsoid, straw-yellow, with a thickened wall. *Basidia* 20–25 x 3–3.5 µm, cylindrico-ventricose, with eight, short sterigmata. *Peridiopellis* of thin-walled hyphae, 1.5–5 µm diam., often inflated, with clamp-connexions.

Habitat: restricted to sandy soil, mainly sand-dunes.

Distribution and frequency: locally common in Britain in suitable dune systems. More common in southern Europe where it can occur in other habitats.

Other remarks: the dentate pileal disk margin has led to the common name 'Toothed Phallus'. Much has been made of the less foetid odour of this species, Rea (1922) suggesting that the smell of the species, as *P. imperialis*, was 'pleasant, like that of liquorice', whilst Phillips & Plowright (1876) thought the epithet *iosmus* to be appropriate, reporting a 'pleasant violet odour, not offensive until the third day after gathering'.

Dune Stinkhorn

Fig. 139. *Phallus hadriani*. (Spain, Madrid, 1 Oct. 1989, Gomez-Ferreras, MA-Fungi 31654).

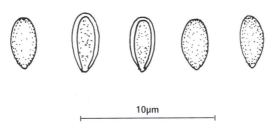

Fig. 140. *Phallus hadriani* (Lincolnshire, Gibralter Point, March 1966, Waterhouse). Spores.

British Puffballs, Earthstars and Stinkhorns

MUTINUS Fr.,
Summ. Veg. Scand. 2: 434 (1849).

Cynophallus (Fr.) Corda, *Icon. Fung.* 6: 14 (1854).
Corynites Berk. & Curtis in *Trans. Linn. Soc.* 21: 149 (1855).
Caromyxa Mont., *Syll. Cryptog.* : 281 (1856).

Immature fruitbody hypogeous, ovoid to pyriform. *Peridium* whitish, membranous, smooth, stratified with mucilaginous middle layer, becoming volvate with apical lobes at maturity. *Gleba* borne subapically over surface of receptacle, dark olivaceous brown, with weak to strongly foetid odour. *Pileus* not developed. *Receptacle* stipe-like, slender, apically tapering, cylindrical, hollow, pseudoparenchymatous, pale coloured, perforate or not. *Indusium* absent. *Spore mass* olive-brown. *Spores* small, ellipsoid, pale yellowish. *Habitat* in damp woodlands, rarely attached to rotting wood. *Type species: Mutinus caninus* (Huds.: Pers.) Fr.

Key to British Species

1. Lower receptacle pale pinkish to orange; gleba rather persistent; odour very mild; fairly common .... **1. Dog Stinkhorn** (*M. caninus*)
1. Lower receptacle carmine-red; gleba soon removed; odour strongly foetid; rare .... **2. Red Stinkhorn** (*M. ravenelii*)

1. DOG STINKHORN

Map No. 68

Mutinus caninus (Huds.: Pers.) Fr., *Summ. Veg. Scand.* 2: 434 (1849).
Phallus caninus Huds.: Pers., *Synops. Meth. Fung.*: 245 (1801); Huds., *Fl. Angl.* Edit.2, 2: 630 (1778).
Phallus inodorus Sow., *Col. Fig. Engl. Fung.* 3: pl. 330 (1801).

Selected descriptions: Massee (1889); Pilát 1958: 50); Rea (1922: 23, as *Cynophallus*)

Selected illustrations: Breitenbach & Kränzlin (1986: pl.526); Gerhardt (1985: 216); Michael, Hennig & Kreisel (1986: pl. 141a); Phillips (1981: 256); Sarasini (1992: pl. 3–5); Sowerby (1801: pl. 330); Wakefield & Dennis (1981: 102, pl. 108/2).

Diagnostic characters: slender, cylindrical, pinkish orange receptacle; glandiform apex with olive-brown gleba and weak odour.

Peridium 2–4 high x 2–2.5 cm, elongate ovoid or pyriform ('witches egg'), initially hypogeous then epigeous, white or yellowish, membranous, smooth, with a central mucilaginous layer, basally attached by a white rhizomorph; finally splitting to become volvate with 2–3 apical lobes. *Receptacle* stipitiform, 6–12 cm tall, 1–1.5 cm diam., cylindrical or tapering above, apically perforate or not, at first white soon pinkish to pale orange, brick-red to deep orange just below the gleba, with a spongy structure, hollow; sterile zone inferior and demarcated from the glandiform, orange-red, glebiferous region by a swollen, ring-like zone. *Gleba* covering and confined to the upper receptacle, mucilaginous, olive-brown, smooth and shiny, with only a weakly foetid odour. *Spore mass* olive-brown.
  Basidiospores 4.5–6.5 x 1.8–3 μm.

Habitat: on soil in damp, usually deciduous woodland, often around decaying stumps, rarely on rotting wood.

Distribution and frequency: occasional to fairly common in Britain and throughout the northern hemisphere, from July to October.

Other remarks: originally described by Hudson (1778) from near Shrewsbury in Shropshire, and illustrated by Sowerby (1801) as a new species, *Phallus inodorus*.

British Puffballs, Earthstars and Stinkhorns

RED STINKHORN

Map No. 69

Mutinus ravenelii (Berk. & Curtis) E. Fisch. in Sacc., *Syll. Fung.* 7: 13 (1888).
Corynites ravenelii Berk. & Curtis apud Berk. in *Trans. Linn. Soc.* 21: 151 (1853).
M. bambusinus sensu Cooke in *Grevillea* 17: 17 (1888) non Zollinger, *Syst. Verz. Ind. Arch. Jahr. 1842–48*, Heft 1: 11 (1854).

Selected descriptions: Berkeley (1853:151); Cooke (1888: 17)

Selected illustrations: Berkeley (1853: pl.19/4); Cooke (1888: pl.173f–1); Knudsen (1986: 64, fig.1); Michael, Hennig & Kreisel (1986: pl.141b).

Diagnostic characters: slender, cylindrical, carmine red receptacle; obtusely acute , often perforate apex.

Peridium 1.8–2 cm high, 1.5–2 cm diam., ovoid ('witches egg'), hypogeous then epigeous, white or with greyish tints, membranous, with a mucilaginous central layer, basally attached, finally splitting and becoming volvate, apically lobed. *Receptacle* stipe-like, 6–8.5 cm high, relatively thickened, around 1–1.8 cm diam., cylindrical, often apically perforate, characteristically carmine-red, spongy, hollow; apical glebiferous region more often obtusely pointed but variable, tapering to clavate. *Gleba* confined to upper region of receptacle, dark olivaceous brown, mucilaginous but not persistent, smooth and shiny, with a very strong foetid odour. *Spore mass* olive-brown.
  Basidiospores 5–7 x 1.8–2.5 µm.

Habitat: damp deciduous woodland and gardens.

Distribution: Rare, only found in Britain in S. E. England. First reported in continental Europe from Berlin in 1943 (Ulbrich 1943), but now widespread (Kosonen 1986). Common in North America.

Other remarks: Originally described from North America, as the only species of *Corynites* Berk. & Curtis. It has been erroneously cited several times in the British literature (Cooke, 1888: 17, pl.173; Massee 1889: 89; Ramsbottom 1953: 185; Rea 1922: 23) as *M. bambusinus*, a species described from Java. The original British record (as *M. bambusinus*) was based on an 1888 collection from nurseries in Windlesham, Sunningdale, Surrey, growing under glass with *Arundinaria*, where it occurred for forty years. *Mutinus bambusinus* has a purplish, upper, acutely tapering receptacle, and is tropical.
  Mutinus ravenelii is similar to *M. caninus* but has a carmine-red receptacle and a more strongly foetid odour.

178

Red Stinkhorn

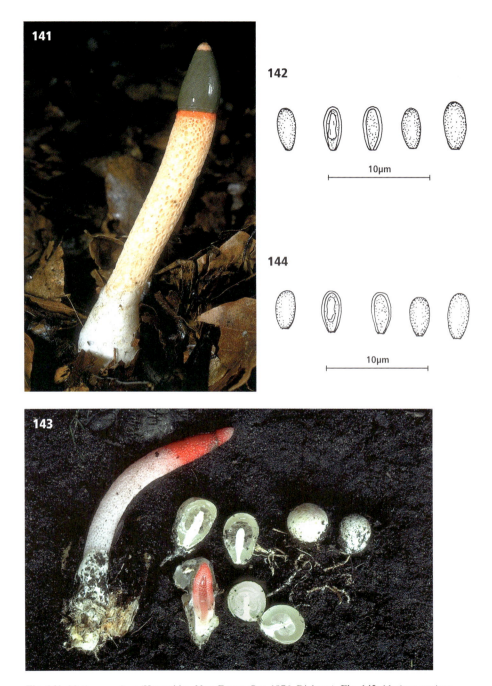

Fig. 141. *Mutinus caninus* (Hampshire, New Forest, Oct. 1976, Dickson). Fig. 142. *Mutinus caninus* (Dunbarton, Sept. 1963, Dring). Spores. Fig. 143. *Mutinus ravenelii* (Denmark, Skåne, Nitare). Fig. 144. *Mutinus ravenelii* (Surrey, Windlesham, Sunningdale Nurseries, April 1929, White). Spores.

CLATHRACEAE E. Fisch.
in Engler & Prantl, *Nat. Pflanzenfam.* 1 (1**): 280 (1900).

Immature fruitbody ('myco-egg') globose to ovoid, comprising a smooth peridium enclosing the gleba and unexpanded receptacle; receptacle connected to peridium by sutures. *Mature fruitbody* multipileate, with the gleba borne on inner (adaxial), sometimes lateral, surfaces of a net-like to stellate receptacle. *Peridium* splitting at maturity but retained as a basal volva. *Gleba* becoming mucilaginous, deliquescent, olivaceous brown, foetid. *Receptacle* well developed, tubular but often anastomosing to become spongy, forming columns variously united into a network or diverging branches, with or without a stem-like base. *Spore mass* olive-brown.

Basidiospores very small, cylindrical, subhyaline to pale brown, smooth, with a slightly thickened wall. *Basidia* 6- or 8-spored. *Peridiopellis* three-layered, with a middle gelatinous stratum. *Type genus*: *Clathrus* Mich.: Pers.

As in the case of *Phallaceae*, the clathroid species expand and grow very rapidly following apical rupture of the peridium. They prefer warmer localities and, although some are now locally common, they have a restricted distribution in Britain and most are only rarely encountered. Of the five species recorded from Britain, all are assumed to be of exotic origin, although *C. ruber* could be native in south-east England. A detailed monographic account of *Clathraceae* was published by Dring (1980), who regarded their evolutionary development not as a simple linear sequence but as a series of parallel phyletic lines. There is no indication of a phyletic relationship outside the *Phallales*, and perhaps the earliest ancestry might be looked for around the *Hysterangiaceae* and *Protubera* Möller. The receptacle is obviously adapted to assist spore dissemination by insects, often resembling a colourful flower. There is considerable variation of form which can extend to species level, sometimes making identification difficult.

Key to British Genera

1. Receptacle with a well developed pseudostipe, surmounted by radial arms, with gleba borne on upper surface of divergent arms 2
1. Receptacle reticulate to stellate, sessile or with a short pseudostipe bearing the gleba internally 3

    2. Receptacle with a cylindrical, sterile stipe, 4–7 cm tall, surmounted by 4–7 short, conical arms, initially apically fused, soon free and slightly curving outwards; whitish below to pink, reddish or orange above; with a persistent, membranous peridial volva .... **2. Lizard's Claw** (*Lysurus cruciatus*)

    2. Receptacle with a cylindrical, fusoid or short tapering pseudostipe with a broad apical disk with numerous (5–22), thin, tentacular arms arranged around the margin; pale pink .... **3. Starfish Fungus** (*Aseroe rubra*)

3. White, lattice-like receptacle, usually more than 10 cm diam., with no basal-apical differentiation, becoming detached from remnants of peridial volva; arms simple tubular .... **4. Basket Fungus** (*Ileodictyon cibarium*)
3. Salmon pink to blood-red receptacle with basal-apical differentiation; remaining attached to peridial volva; arms complex, spongy .... **1. Cage Fungi** (*Clathrus*)

1. CAGE FUNGI

Clathrus Mich.: Pers., *Synops. Meth. Fung.* 2: 241 (1801); Mich, *Nov. Gen. Pl.*: 214 (1729).
Aserophallus Lepr. & Mont. in *Ann. Sci. Nat., Bot.* sér.3, 4: 360 (1845).
Anthurus Kalchbr. & MacOwan apud Kalchbr. & Cooke in *Grevillea* 9: 2 (1880).
Linderia G. Cunn. in *Proc. Linn. Soc. N.S.W.* 56: 192 (1931).

Immature fruitbody sub-hypogeous, 2–4 cm diam., more or less globose, consisting of an outer, pale peridium, enclosing an unexpanded receptacle embedded in a gelatinous endoperidium, with basal mycelial strands. *Peridium* off-white to ochre-yellow, membranous, with a gelatinous, translucent yellow endoperidium; position of internal peridial sutures marked by reticulations on the surface of unexpanded fruitbody; rupture apical into irregular lobes at maturity, the peridium retained as a volva to which the receptacle remains attached. *Gleba* mucilaginous, dark greenish, deliquescent at maturity, borne on the inner surfaces of the receptacle either dispersed over entire surface or restricted to numerous, scattered 'glebifers'. *Receptacle* fragile, rapidly expanding on release, bright red in British species, paler at base, always showing some apico-basal differentiation, either forming a latticed, hollow sphere (clathroid) or stellate, with 4–6 radiating arms (anthuroid); arms with spongy texture, hollow, complex, consisting of many intercommunicating tubes. *Odour* finally strongly foetid. *Spore mass* olive-brown.

Basidiospores ellipsoid, hyaline to straw-yellow, smooth. *Habitat* solitary or gregarious, on rich soil or rotting wood in woodland and gardens, preferring warmer localities. *Type species: Clathrus ruber* Mich.: Pers.

Dring (1980) considered *Clathrus* to be the most primitive genus of the family and the most variable, from which several evolutionary series may be derived, principally the clathroid-series, with a spherical latticed receptacle, and the anthuroid-series, with stellate arms. The latter has been considered, until recent years, as a separate genus, *Anthurus*, but the characters of the receptacle arms being apically united in the early stages, together with the gleba covering the inner surfaces, indicates that a continous developmental series exists and no generic delimitation can be maintained.

Key to British Species

1. Receptacle stellate with 4–8 reflexed, radiating arms, 3–7 cm long, initially fused at apex; white at base to deep pink above, inner glebal surface blood-red
   .... **1. Devil's Fingers** (*C. archeri*)
1. Receptacle lattice-like, up to 12 cm diam., with isodiametric mesh, usually more elongated towards base; salmon-pink, with scarlet-red inner glebal surface
   .... **2. Red Cage Fungus** (*C. ruber*)

1. DEVIL'S FINGERS

Map No. 70

Clathrus archeri (Berk.) Dring in *Kew Bull.* 35: 29 (1980).
Lysurus archeri Berk. apud Hooker, *Fl. Tasm.* 2: 264 (1860).
Anthurus archeri (Berk.) E. Fisch. in *Jahrb. K. Bot. Gart., Berlin* 4: 81 (1886).
Anthurus aseroeformis (E. Fisch.) McAlpine in Lloyd, Mycol. Notes 31 in *Mycol. Writ.* 2: 408 (1908).

Selected descriptions: Cunningham (1944: 101); Dennis & Wakefield (1946: 141); Dring (1980: 28, fig. 8); Pilat (1958: 80, fig.17); Ramsbottom (1953: 189).

Selected illustrations: Breitenbach & Kränzlin (1986: pl.523); Fuhrer (1985: 102); Gerhardt (1985: 214); Marchand (1976: pl.164); Michael, Hennig & Kreisel (1986: pl.142); Rinaldi & Tyndalo (1972: 224, fig. 1); Vesterholt & Sørensen (1989: 16).

Diagnostic characters: fruitbodies often clustered; peridium white to brown, furrowed, furfuraceous; 4–8 blood-red radiating arms; short stem attached to volva.

Immature fruitbody 2–4 cm diam., up to 6 cm high but normally much less, globose to ovoid, with a white, mycelial strand. *Peridium* white to ochre-buff, furfuraceous, marked with furrows corresponding to the receptacle arms, with a thin, greenish brown gelatinous endoperidium. *Gleba* olivaceous brown then black, borne on inner surfaces of receptacle, mucilaginous, drying as a broken reticulum over the surface. *Receptacle* comprising a variable, short, hollow 'stem', 3–5 (–7) cm long, 1–2.5 cm diam., pale below becoming pinkish red above, tapering with abruptly slender basal attachment, surmounted by 4–7 (–8), pointed, chambered, pinkish orange arms, (3–) 5–7 (–18) cm long, apically united on initial expansion but usually breaking away and spreading horizontally and drooping; inner surface blood-red, transversely rugulose. *Spore mass* olivaceous brown.
Basidiospores 4.5–8.5 x 2.3–3.5 μm, ellipso-cylindrical, straw-yellow, smooth, with a thickened wall. *Basidia* 6-spored.

Habitat: on soil, decaying wood chips or stumps, August to November.

Distribution: very rare but can occur in large numbers and repeatedly in the same locality; now spreading in southern England, possibly as a result of an increased use of wood chippings as mulch. An introduced species. Originally described from Tasmania, and listed as not uncommon in Australia and New Zealand by Cunningham (1944). First recorded from Britain (Penzance, Cornwall) by Dennis & Wakefield (1946), and now known from three mainland centres (Cornwall, Sussex and Kent), and from Guernsey. Ramsbottom (1953), Herrmann (1962) and Kriegelsteiner (1992) have discussed its distribution and spread throughout Europe, where it was first recorded in Vosges, France, in 1914.

Other remarks: The long receptacle arms and gleba extending down the inner surface of the 'stem' are distinctive. The number of arms is variable and they occasionally bifurcate or anastomose. Many of the early British reports of *Aseroe rubra* Labillardière should be referred here.

Devil's Fingers

Fig. 145. *Clathrus archeri* (Cornwall, Penzance, July 1988, Dickson).

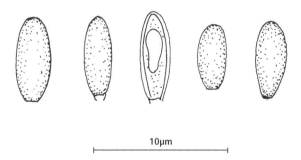

Fig. 146. *Clathrus archeri* (East Sussex, October 1988, Wells). Spores.

2. RED CAGE FUNGUS

Map No. 71

Clathrus ruber Mich. : Pers., *Synops. Meth. Fung.* 2: 241 (1801); Mich. *Nov. Pl. Gen.*: 214 (1729).
C. cancellatus Tourn. ex Fr., *Syst. Mycol.* 2: 288 (1823).

Selected descriptions: Dring (1980: 14); Marchand (1976: 162); Pegler (1990: 170); Rea (1922: 21)

Selected illustrations: Breitenbach & Kränzlin (1986: pl.524); Dennis, Reid & Spooner (1977: fig.4J); Gerhardt (1985: 213); Marchand (1976: pl.379); Massee (1889: pl.3/46); Michael, Hennig & Kreisel (1986: pl.140a); Ramsbottom (1958: pl.IX, 37b)

Diagnostic characteristics: receptacle lattice-like, red; spongy arms forming a mesh; receptacle remaining attached to volva.

Immature fruitbody sub-hypogeous becoming epigeous, 3–6 cm diam., subglobose, apically rupturing into irregular lobes, containing an unexpanded receptacle attached to the peridium by white sutures and embedded in a pale yellowish, mucilaginous endoperidium; attached by a thick, basal mycelial cord. *Peridium* off-white to greyish ochre, with a grooved reticulum over the surface; gelatinous endoperidium up to 3 mm thick. *Receptacle* fragile, sessile, 10–12 cm high, 7–9 cm diam., salmon-pink to scarlet-red, paler towards the base, expanding to a hollow, lattice-like sphere of anastomosing arms, forming large polygonal meshes (c. 30), more elongated towards the base; arms about 1.5 cm thick, triangular in section, spongy with a large central tube and several smaller lateral tubes, with flattened but rugulose outer surface; remaining attached to remnants of basal peridial volva. *Gleba* olivaceous yellow drying to black, mucilaginous, at first covering the inner surfaces of the receptacle, finally reduced to small, scattered areas. *Odour* very foetid.
Basidiospores 4–6 x 1.5–3.2 μm, ellipso-cylindrical, hyaline to pale greenish yellow, smooth, thin-walled. *Basidia* 6-spored.

Habitat: solitary or in small groups, at edge of woodland and in parkland; amongst leaf-litter, under hedges, preferring warm localities.

Distribution: rare in Britain, restricted to southern England (south of line Wash to Dyfed), apart from Dublin and three records in southern Scotland. Bromfield (1843) cited the first British records from several localities on the Isle of Wight. Native to Mediterranean regions, extending into central and northern Europe.

Other remarks: a beautiful species; the only clathroid species of *Clathrus* known in Europe, although 16 species of the genus are known worldwide, most of which are tropical. Dennis (1955) regarded *C. ruber* as alien to Britain since almost all records were from pleasure grounds or gardens.

Red Cage Fungus

Fig. 147. *Clathrus ruber* (Kent, 1992, Laessøe).

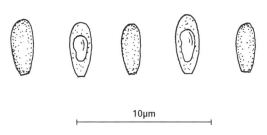

Fig. 148. *Clathrus ruber* (Jersey, Aug. 1963, Nicolle). Spores.

2. LIZARD'S CLAW

Lysurus Fr.,
Syst. Mycol. 2: 285 (1823).

Receptacle comprising a relatively long stipe, and a differentiated upper fertile region in the form of obtuse, vertical arms. 5 species worldwide. *Type species*: *Lysurus mokusin* Fr.

Lysurus cruciatus (Lepr. & Mont.) Lloyd, *Mycol. Writ.* 3, *Synops. Known Phall.*: 40 (1909).

Map No. 72

Aserophallus cruciatus Lepr. & Mont. in *Ann. Sci. Nat., Bot.* sér. 3, 4: 36 (1845).
Anthurus cruciatus (Lepr. & Mont.) E. Fisch. in *Denkschr. Schweiz. Ges. Nat.* 36: 41 (1900).
Misapplied name: *Lysurus gardneri* sensu Ramsbottom (1953), Palmer (1968), Pilát (1958) non *L. gardneri* Berk. in *Hooker, Lond. J. Bot.* 5: 535 (1846).

Selected descriptions: Cunningham (1944: 105); Dring (1980: 76, fig. 24 C–F); Pilát (1958: 713, fig.20).

Selected illustrations: Michael, Hennig & Kreisel (1986: pl. 140b); Murrill (1912: 167, pl. 68/8); Ramsbottom (1953: pl. XIb); Rea (1904: 57, pl. 3A).

Diagnostic characteristics: columnar receptacle with 4–7 short, apical arms; gleba restricted to inner surface of arms; persistent volva.

Immature fruitbody subhypogeous, globose to obovoid, up to 5 cm diam., attached by a white mycelial cord, comprising an outer peridium which ruptures irregularly, containing an unexpanded receptacle. *Peridium* white, membranous, with meridional grooves, retained as a persistent volva. *Gleba* copious, yellowish brown, deliquescing slowly, with a slightly foetid odour, covering inner surface of receptacle arms. *Receptacle* off-white, with a cylindrical, sterile stipe, 4–10 cm tall, about 1.5–2 cm diam., whitish to cream colour below but often becoming pink, reddish or orange above, spongy, surmounted by 4–7 short, conical, tapering arms, 1.5–2 (–4) cm long, initially apically fused soon free and slightly curving outwards; outer surface concave, furrowed, smooth, white, orange-red to reddish brown; inner surface transversely wrinkled; remaining attached to peridial volva.
Basidiospores 4.5–6 x 1.5–2.3 µm, ellipso-cylindrical, subhyaline, smooth, thin-walled.

Habitat: on rich or highly manured soil or rotting hay.

Distribution and frequency: common in Australia and New Zealand, but rare and presumably introduced in Britain. First recorded from Europe (Germany) in 1902, and Rea (1904) described a collection from Kidderminster. Park (1981) reported the species from Lisbellaw, Co. Fermanagh, Republic of Ireland.

Other remarks: more often reported under the name *L. gardneri*, which is a south-east Asiatic species in which the arms have a sterile narrow base and the fertile portion is expanded into gill-like processes. The presence of reddish tints in *L. cruciatus* is very variable and sometimes absent, which has resulted in an extensive synonymy (see Pilát 1958 : 85).

186

Lizard's Claw

Fig. 149. *Lysurus cruciatus* (Spain, Seville, 17 Jan. 1987, Martinez, MA-Fungi 22359).

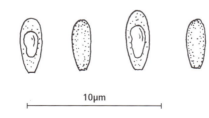

Fig. 150. *Lysurus cruciatus* (London, Chiswick Park, 1916). Spores.

3. STARFISH FUNGUS

Aseroe Labill.,
Nov. Holl. Pl. Spec. 2: 124 (1806).

Receptacle comprising a pseudostipe with an apical disk, from the margin of which arise numerous, long, acute, paired arms. *Gleba* borne on disk and adaxial surfaces only. 2 species worldwide. *Type species*: *Aseroe rubra* Labill.

Aseroe rubra Labill., *Nov. Holl. Pl. Spec.* 2: 124 (1806).

Map No. 73

A. pentactina Endl., *Icon. Gen. Pl.* 1: 50 (1838).
A. viridis Berk. in *Hooker, Lond. J. Bot.* 5: 535 (1844).
A. actinobola Corda, *Icon. Fung.* 6: 23 (1854).
A. hookeri Berk., *Fl. New Zeal.* 2: 187 (1855).

Selected descriptions: Cunningham (1944: 107); Dring (1980: 81, fig. 25); Petch (1908: 175); Pilát (1958: 88, fig.21)

Selected illustrations: Fuhrer (1985: 102–103); Massee (1897: p.115)

Diagnostic characters: broad disk at pseudostipe apex; numerous, slender bifurcating 'tentacles'; pink pseudostipe.

Immature fruitbody subhypogeous, up to 3 cm diam., subisodiametric, obovoid to somewhat turbinate, with basal mycelial strands. *Peridium* off-white, creamy white or pale brown, membranous, at times tessellated, with pale grey, gelatinous endoperidium. *Gleba* bright red, covered by a brown slime. *Receptacle* forming a pseudostipe, 5–9 cm tall, 1–2.5 cm diam. at apex, cylindrical, subfusoid or tapering below, pale pink, whitish at base, spongy, hollow; apex flattened to form a disk (–3.5 cm diam.) bearing 5–11, chambered, narrow, tapering arms, which regularly bifurcate about 1.5 cm from their base, 1.5–3.5 cm long, 1–3 mm thick, strawberry-red, fully expanded at maturity. *Spore mass* greyish to olive-brown. *Odour* unpleasant, foetid but weak.

Basidiospores 4.5–7 x 1.7–2.5 µm, ellipso-cylindrical, truncate at base, pale brown, smooth, with a slightly thickened wall. *Basidia* not seen.

Habit: typically found in subalpine grasslands and open forest, generally on rich soil in sheltered places. British record in leaf litter on acid, sandy soil.

Distribution and frequency: extremely rare; only known from the Royal Botanic Gardens, Kew, where in 1829 it fruited in a greenhouse on soil imported from Australia, and from a recent collection (1992) from Oxshott Heath, Surrey, on sandy soil amongst forest litter. An introduced species, typically southern tropical to subtemperate and subalpine.

Other remarks: A variable species with regard to size, length of pseudostipe, and number of arms. Distinguished from *Clathrus* and *Lysurus* by the slender, bifurcating arms and the flattened apical disk of the receptacle. *Clathrus* lacks a disk and the gleba extends down the inner surface of the pseudostipe. Numerous varieties have been described but these tend to intergrade.

Starfish Fungus

Fig. 151. *Aseroe rubra*. (Surrey, Esher, Oct. 1993, Spooner).

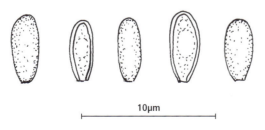

Fig. 152. *Aseroe rubra* (Surrey, Esher, Oct. 1993, Spooner). Spores.

4. BASKET FUNGUS

Ileodictyon Tul. in Raol
in *Ann. Sci. Nat., Bot.* sér. 3, 2: 114 (1844).

Receptacle clathroid, radially symmetrical, becoming detached from volva; *arms* formed of one to few continuous tubes. 2 species worldwide. *Type species*: *Ileodictyon cibarium* Tul.

Ileodictyon cibarium Tul. apud Raol in *Ann. Sci. Nat., Bot.* sér.3, 2: 114 (1844).

Map No. 74

Clathrus cibarius (Tul.) E. Fisch. in *Jahrb. K. Bot. Gart., Berlin* 4: 74 (1826).

Selected descriptions: Cunningham (1944: 110, pl.XII–XIII); Dring (1980: 56, fig. 17); Reid & Dring (1964: 293).

Selected illustrations: Fuhrer (1985: 101 & 104); Reid & Dring (1964: fig.1).

Diagnostic characters: receptacle globose, white with radially symmetrical lattice; soon detached from volva; receptacle arms tubular, not spongy.

Immature fruitbody subhypogeous, subglobose to ovoid, up to 7 cm diam., consisting of an unexpanded receptacle with characteristic concertina-like folding, within a membranous peridium, attached by branching, white mycelial cords. *Peridium* off-white, with reticulate grooves, rupturing irregularly at the apex. *Gleba* olivaceous brown, mucilaginous, covering inner surfaces of receptacle, sometimes finally hardening and contracting to restricted zones. *Receptacle* sessile, pure white, translucent, more or less globose but becoming flaccid, 8–10 (–25) cm diam., composed of a hollow sphere of anastomosing arms forming isodiametric, polygonal meshes (c. 10–30), showing a radial symmetry with no apico-basal differentiation; arms c. 1 cm diam., not thickened at intersections, hollow, simple tubular; detached from the peridial volva. *Odour* foetid or recalling Camembert cheese.
  Basidiospores 4.5–7 x 2–2.8 μm, ellipso-cylindrical, subhyaline with a greenish brown tint, smooth, with a thickened wall.

Habitat: at the edge of woodland clearings or on disturbed soil.

Distribution and frequency: very rare and introduced, usually occurring in winter months. First British record from Hampton, Middlesex, in 1955 (Reid 1985), with several later collections found within a 5 km radius, in Middlesex (Hampton, Heston, Hounslow) and Surrey (Thames Ditton), all in garden localities. An apparently unrelated find, also in a garden, is known from near Ipswich, Suffolk. Not recorded elsewhere in Europe. Extremely common in New Zealand, less so in Australia (Cunningham, 1944)

Other remarks: the fruitbody may be adapted to tumbling in the wind to aid dispersal and distribution. Expansion can be extremely rapid once the receptacle is released, and a detailed account of this phenomenon is provided by Colenso (1893).

Basket Fungus

Fig. 153. *Ileodictyon cibarium* (Middlesex, Hounslow, Feb. 1977, Panter).

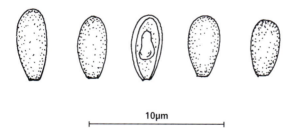

Fig. 154. *Ileodictyon cibarium*. (Middlesex, Hampton, Dec. 1984, Russell). Spores.

HYPOGEOUS GASTEROID FUNGI

The hypogeous gasteroid fungi are considered in detail in Pegler, D. N., Spooner, B. M. & Young, T. W. K. (1993). *British Truffles. A Revision of the British Hypogeous Fungi.* Modified versions of the keys to families, genera and species are provided below:

Key to Orders and Families

1. Spores hyaline or nearly so 2
1. Spores yellowish brown, ferruginous, fuscous brown or black 6

    2. Spores ornamented, with an amyloid myxosporium; peridial context heteromerous with sphaerocytes; clamp-connexions absent
    .... **Milk Cap Truffles, Russulales, Elasmomycetaceae**
    2. Spores inamyloid, smooth or ornamented; peridial context homoiomerous, lacking sphaerocytes; clamp-connexions present or absent 3

3. Spores ornamented, with or without a myxosporium, globose or nearly so; basidia 2–4 spored .... **Agaricales**, 4
3. Spores smooth, with a slightly thickened wall, sometimes with a loose myxosporium; basidia 4–8 spored 5

    4. Spores thick-walled, with a reticulato-alveolate ornamentation overlain by a thick, gelatinous myxosporium; glebal chambers gelatinized; clavato-pedunculate basidia autolysing; clamp-connexions absent; columella absent
    .... **Honeycomb Truffle, *Leucogaster nudus*** (Hazl.) Hollos
    4. Spores thin-walled, with an ornamentation of isolated conical spines, lacking a myxosporium; glebal chambers empty to partially filled with spores; euhymenial; clamp-connexions present; columella absent
    .... **Flesh-Pink Truffle, Hydnangiaceae, *Hydnangium carneum*** Wallr.

5. Dendroid columella present; spores narrowly fusoid, often with a wrinkled, separating myxosporium; clamp-connexions present; rhizomorphs present
    .... **Stinkhorn Truffles, Phallales, Hysteriangiaceae**
5. Columella absent; spores narrowly ellipsoid, basally truncated and less than 10 μm long; clamp-connexions absent; fruitbodies relatively large, often covered with conspicuous fine rhizomorphs; always associated with *Pinaceae*
    .... **Beard Truffles, Boletales, Rhizopogonaceae, *Rhizopogon*** Fr.

    6. Spores with 8–16, prominent, meridional, solid costae, often anastomosing, fuscous brown; peridium soon evanescent, revealing a labyrinthoid gleba; clamp-connexions present; columella present, soon reduced to a sterile base
    .... **Boletales, Gautieriaceae, *Gautieria morchelliformis*** Vittad.
    6. Spores not longitudinally costate 7

7. Spores dark olive-brown, smooth, with a truncate base; gleba black, solid, with gelatinized chambers
    .... **Slime Truffles, Agaricales, Melanogastraceae*, Melanogaster*** Corda
7. Spores yellowish brown, subhyaline to ferruginous; columella absent 8

193

British Puffballs, Earthstars and Stinkhorns

8. Spores subglobose to globose, 9–12 µm diam., pale yellowish brown 10

8. Spores either ellipsoid to fusoid, 10–20 µm long, smooth or verruculose, or very small, 3–5 µm diam., lacking a basal corona, often with a loose myxosporium, yellowish to rusty brown, non-dextrinoid 9

9. Peridiopellis an epithelium; spores 14–19 µm long, mucronate, verruculose, with a . loosely applied myxosporium
.... **Cortinariales, Hymenangiaceae,** *Hymenangium album* Klotzsch

9. Peridiopellis a repent epicutis; spores mucronate or not, either smooth but often with wrinkled myxosporium; or globose and verruculose
.... **Cortinariales, Hymenogasteraceae**

10. Spores irregularly spinose and with a peri-appendicular corona; fruitbody bright red .... **Carrot-Red truffle, Stereales, Stephanosporaceae,** *Stephanospora caroticolor* (Berk.) Pat.

10. Spores coarsely ornamented, with tuberculate verrucae, lacking a basal corona; fruitbody whitish .... **Leaf-litter truffles, Cortinariales, Octavianinaceae**

Key to British species of Elasmomycetaceae Kreisel

1. Context containing laticiferous elements, producing a white latex; fruitbody 1.5–3 cm diam., orange-red to dark reddish brown; tramal plates lacking sphaerocytes; spores 12.5–16 x 10–12.5 µm, isolated verrucae; alliaceous odour present
.... *Zelleromyces stephensii* (Berk.) A.H.Smith

1. Context not producing a latex; sphaerocytes present and abundant in tramal plates; spores 10–15 x 10–14 µm, with isolated spines, up to 1.5 µm long
.... *Gymnomyces xanthosporus* (Hawker) A.H.Smith

Key to the genera of Hymenogasteraceae Vittad.

1. Spores small, globose or nearly so, 3–5 µm diam., verruculose, lacking a myxosporium; fruitbody chalky white, with a flaking peridium, pulverulent gleba, and conspicuous mycelial cords; associated with grasses.
.... **Steppe Truffle,** *Gastrosporium simplex* Mattirolo

1. Spores ellipsoid to fusoid, 10–20 µm long, smooth but often enveloped in a strongly wrinkled myxosporium; fruitbody with more persistent, non-powdery peridium, loculate gleba, and inconspicuous basal, mycelial threads; associated with tree roots
.... **Nut Truffles,** *Hymenogaster* Vittad.

Hypogeous Gasteroid Fungi

Key to the British species of *Hymenogaster* Vittad.

1. Spores lacking a persistent myxosporium, 19–23 x 9–12 µm, narrowly citriform, sulphur-yellow; fruitbody 0.5–1.5 cm diam., snow white then yellowish brown; gleba greenish to sulphur-yellow; odour sweetish, recalling vanillin or phenol; common under conifers, occasionally deciduous trees .... **Yellow Nut Truffle,** *H. luteus* Vittad

1. Spores enveloped in a loose,wrinkled or fragmenting myxosporium 2

2. Spores with a pronounced and persistent mucronate apex 3

2. Spores more or less fusoid,with an obtusely rounded apex, never mucronate 7

3. Spores more than 25 µm long, fusoid-citriform 4

3. Spores less than 25 µm long, citriform 6

4. Fruitbody 1–2 (–4) cm diam., soon becoming bright lemon-yellow to golden yellow, finally nigrescent; gleba red-brown to umbrinous; spores 20–40 x 14–19 µm, with a loose myxosporium; odour 'cheese-like' or phenolic; common .... **Lemon-Coloured Nut Truffle,** *H. citrinus* Vittad.

4. Fruitbody not bright yellow, rather off-white discolouring brownish; odour slight 5

5. Spores elongate, 30–47 x 13–19 µm, with a wrinkled myxosporium; fruitbody 0.5–2.5 (–4) cm diam., with olivaceous tints; gleba creamy brown to dull brown; most common species, often in large numbers .... **Olive Nut Truffle,** *H. olivaceus* Vittad.

5. Spores 24–32 x 12–15 µm, citriform, with myxosporium fragmenting into irregular patches; fruitbody 0.5–1 cm diam., often with a basal groove; gleba yellowish brown to golden brown; rare .... *H. sulcatus* Hesse

6. Gleba soon acquiring pale pinkish tints, finally grey-brown; spores 13–17 x 8–12 µm, with myxosporium fragmenting into irregular patches; fruitbody 0.5–2 cm diam., off-white bruising reddish, often with a white rhizomorph; odour fungoid; common .... **Thin Nut Truffle,** *H. tener* Berk. & Br

6. Gleba pale brown to greyish black,variegated; spores 16–22 x 12–17 µm, with a closely applied but very strongly wrinkled myxosporium; fruitbody 0.2–1 cm diam., pure white to greyish; odour none; uncommon .... *H. arenarius* Tul.

7. Spores more than 26 µm long 8

7. Spores 19–26 x 10–14 µm, ellipso-fusoid 9

8. Spores 22–37 x 11–14 µm, fusoid,with a longitudinally wrinkled myxosporium; fruitbody 0.5–1.5 cm diam., whitish to dull yellowish brown; gleba soon darkening to fuscous black; common .... **Common Nut Truffle,** *H. vulgaris* Tul

8. Spores 19–32 x 12–16 µm, with a closely applied, irregularly wrinkled myxosporium; fruitbody 0.5–1.5 cm diam., off-white to very dark brown, peridium often cracking; gleba ferruginous to dark fuscous .... **Cracking Nut Truffle,** *H. muticus* Berk. & Br

9. Spores with a very loose, wrinkled myxosporium, sometimes giving a more or less globose outline; gasterocarp small, less than 5 mm diam., whitish staining brown; gleba brown; known only from type locality .... **Thwaites' Nut Truffle,** *H. thwaitesii* Berk. & Br.

9. Spores with a closely applied, wrinkled myxosporium 10

195

British Puffballs, Earthstars and Stinkhorns

10. Gleba dark greyish brown; spores fuscous brown; fruitbody 0.5–1.5 cm diam., whitish soon cinnamon-brown, with a thick peridium; odour sweet .... ***H. griseus*** Vittad.

10. Gleba dark fuscous brown to black; fruitbody 1–3 cm diam., white to greyish; odour rancid .... ***H. hessei*** Soehner

Key to British species of *Hysterangium* Vittad.

1. Spores more than 15 µm long; fruitbody associated with and often embedded within an extensive white mycelium 2

1. Spores 9.5–12.5 x 3.7–5 µm, cylindrico-subfusoid, mostly with an obtuse apex, overlain by a strongly wrinkled myxosporium; gasterocarp arising from a basal rhizomorph, and not enveloped in an extensive mycelium; peridium thick, coriaceous and brittle when dry; peridiopellis with a broad, pseudoparenchymatous hypodermium .... ***H. coriaceum*** Hesse

2. Spores 12–18 x 5–7.5 µm, ellipso-fusoid with an obtuse apex and a conspicuously wrinkled myxosporium; gleba white to pinkish buff, forming slate-grey to bluish green tints; peridiopellis with a broad hypodermial layer of inflated elements; peridium white to clay-brown, not discolouring on bruising, finally cracking open to reveal the gleba .... **Grey Stinkhorn Truffle,** ***H. nephriticum*** Berk.

2. Spores 17–25 x 6.5–8.5 µm, with an acute or constricted apex, and a tightly appressed myxosporium; gleba white then milk-chocolate-brown, lacking any glaucous tints; peridiopellis lacking a hypodermium; peridium bruising pinkish to rufous brown .... **Thwaites' Stinkhorn Truffle,** ***H. thwaitesii*** Berk. & Br

Key to the British species of *Melanogaster* Corda

1. Spores 6.5–10.5 x 3.5–5 µm, narrowly oblong-ellipsoid; fruitbody reddish ochraceous to fuscous brown; most common species .... **Broome's Slime Truffle,** ***M. broomeianus*** Berk

1. Spores more than 12 µm long 2

2. Spores 14–20 x 8–10.5 (–12) µm, subfusoid to citriform with a subacute to mucronate apex; fruitbody dull olivaceous brown; tramal plates white; rather common, mostly under *Fagus* .... **Stinking Slime Truffle,** ***M. ambiguus*** (Vittad.) Tul

2. Spores 10–14 x 6.5–9 µm, obovoid to broadly ellipsoid, lacking a papillate apex; fruitbody reddish brown; tramal plates pale yellow; very rare .... ***M. intermedius*** (Berk.) Zeller & Dodge

Key to the British species of Octavianinaceae Loquin ex Pegler & Young

1. Spores small, 5–7 µm diam., with low, truncate verrucae, lacking any myxosporium at maturity; fruitbody less than 1 cm diam., white and embedded in a floccose mycelium; gleba greenish yellow, with minute chambers and gelatinous tramal plates, drying hard; clamp-connexions absent .... *Sclerogaster compactus* (Tul.) Sacc.

1. Spores more than 10 µm diam., gleba dark, purplish to blackish; growing in calcareous beech woods 2

    2. Spores 10–14 µm diam., with compound, conical spines (–4µm); gleba whitish to cinnamon-brown, nigrescent, with conspicuous chambers; fruitbody up to 3 cm diam., white then purplish black or greenish; common in western England
    .... *Octavianina asterosperma* (Vittad.) Kuntze

    2. Spores 10–12 µm diam., with verrucae and short ridges (–1.5 µm), with an overlying myxosporium; gleba greyish to dark brown; fruitbody up to 2.5 cm diam., whitish; clamp-connexions present in the peridiopellis
    .... *Wakefieldia macrospora* (Hawker) Hawker

Key to the British species of *Rhizopogon* Fr.

1. Peridium discolouring red to purplish red on bruising; fruitbody 1–3 cm diam.; rhizomorphs scanty 2

1. Peridium not bruising; numerous dark rhizomorphs present; spores less than 8 µm long 3

    2. Spores 7.5–10 x 2.5–4 µm; fruitbody 1.5–3 (–6) cm diam., irregularly ellipsoid, cream-coloured drying reddish brown, with purplish patches; up to 8 cm deep in needle-litter, under *Pinus*
    .... **Blushing Beard Truffle,** *R. roseolus* (Corda) T. M. Fries

    2. Spores 4.5–6.5 x 2.3–3 µm; otherwise similar to *R. roseolus*
    .... **Common Beard Truffle,** *R. vulgaris* (Vittad.) M. Lange

3. Fruitbody white to yellowish brown, not blackening when bruised, with coarse, dark rhizomorphs; spores 5.5–7.5 x 2.5–3 µm, yellowish brown; subepigeal in sandy soil, under *Pinus* .... **Yellow beard Truffle**, *R. luteolus.*

3. Fruitbody whitish, blackening on bruising or exposure, with a surface network of fine rhizomorphs; spores 4.5–7 x 1.5–2.5 µm, narrow, pale brown; under *Picea*, known only from Avon .... **Reticulated Beard Truffle**, *R. reticulatus* Hawker

GLOSSARY

amyloid (of spores and hyphae): staining dark blue with Melzer's reagent (containing iodine).

angiocarpic (of gasterocarp development): spore-producing structures mature whilst retained inside the fruitbody and not exposed.

autolysis (of *Phallales* gleba): breaks down into a liquid at maturity.

ballistosporic (of basidiospore): forcibly discharged from the basidium.

basidiospore: reproductive 'cell', resulting from meiotic division of the nucleus, produced by a basidium.

basidium: the reproductive unit, a single 'cell' in which nuclear fusion and meiosis occur, producing spores, typically four but may be 1–8, at the apex.

capillitium: coarse, thick-walled, hyphae found in the mature gleba of *Lycoperdales, Geastrales, Tulostomatales.*

clamp-connexion (of hyphae): a characteristic structure at the septum, with a curved outgrowth growing out from one segment and by-passing the septum before reuniting with the next segment, allowing transfer of nucleus.

clavate: club-shaped.

columella: a central, sterile structure, positioned vertically in the gleba of certain genera.

coriaceous (of peridium): tough and leather-like.

crenulate (of a margin): having small, rounded teeth.

cupulate (of a fruitbody): cup-shaped.

cyanophilic (of spore and hyphal walls): stains dark blue with cotton blue in lactic acid.

dehiscent (of peridium): splitting to expose the gleba for spore release.

dimitic (of hyphal system): two types of hyphae present.

echinulate: covered with very small, fine spines (echinules).

epigeous (of development): growing or developing at ground level.

epiphragm (in *Nidulariales*): apical region of peridium, which eventually detaches to reveal the mature peridioles.

fascicle: (of hyphae): a cluster.

funiculus (in *Nidulariales*): fine, twisted cord, within a sheath, connecting a peridiole to peridial wall.

gasterocarp: fruitbody of a 'gasteromycete', typically with a sterile peridium enclosing the fertile gleba.

geastroid (of fruitbody): resembling *Geastrum* species, or earthstar.

generative hyphae: basic hyphal type from which all structures are produced, usually thin-walled, branched, septate and actively growing.

glabrous (of a surface): smooth, neither hairy nor scaly.

glandiform: acorn-shaped.

gleba: the spore-producing region found within an angiocarpic fruitbody.

glebifers (in *Clathrales*): small, spore-producing regions, borne on inner surface of the receptacle.

globose: ball-shaped.

gregarious (of fruitbodies): growing as a group but not connected to each other.

hymenium: well defined layer of parallel arranged basidia.

hypha (-e): individual filament which forms the basic structure of both the mycelium and gasterocarp.

hypogeous (of fruitbody): growing or developing below ground level.

inamyloid: negative reaction to Melzer's reagent (see amyloid).

indusium (of *Phallus*): a membranous, spreading veil, formed on the receptacle below the pileus.

infundibuliform: funnel-shaped.

lenticular: lens-shaped, flattened but both surfaces convex.

micron (µm): one-thousandth of one millimetre.

monomitic (of hyphal systems): all hyphae of same type i.e. generative hyphae.

mucilaginous: containing or forming a viscous fluid.

multipileate (of development): several pilei arising from a single stalk.

mycelium: mass of hyphae forming the vegetative thallus.

mycorrhizal: the hyphae of a fungus forming a mutualistic relationship with the roots of a plant or tree; the relationship may be facultative or obligatory.

myxosporium (of a spore): outermost layer of spore-wall, usually sticky, and may form part of the surface ornament.

operculum: a cover or lid.

orthotropic (of basidiospores): manner of development, resulting in a radial symmetry of the spore.

ostiole (of gasterocarp): defined small opening on upper surface.

peridiole (in *Nidulariales*): small, separate body with outer sterile wall and containing part of the gleba with spores.

peridiopellis: outer layer of the peridium.

peridium: sterile, protective wall of gasterocarp, enveloping the gleba.

peristome: well delimited edge around the ostiole.

pileate: bearing or forming a pileus

plectobasidium: a basidium which arises irregularly, not collectively forming a hymenium.

pseudohymenium (in *Nidulariales*): an indefinite fertile layer, formed by plectobasidia.

pseudoparenchymatous: describing a tissue of closely compacted hyphae, appearing cellular when seen in section.

pseudostipe: stem-like base which is not structurally distinct from the upper part of gasterocarp.

pulverulent: powdery.

pyriform: pear-shaped.

receptacle (in *Phallales*): stem-like, net-like or stellate outgrowth which support the gleba.

reniform: kidney-shaped.

reticulate: forming a net.

saprophyte: growing on dead organic matter.

squamose: forming scales.

Glossary

statismosporic (of basidiospore): not forcibly discharged or released from the basidium at maturity.
stellate: star-like.
stipe: structurally distinct stem.

tomentose: having a covering of dense, fine hairs.
tramal plates: layers of sterile tissue, which give rise to the hymenia and basidia within the gleba.
trophocysts (in *Sclerodermatales*): nutritional 'cells' within the gleba, which stimulate young basidiospores to maturity; also called 'nurse-cells'
tuberous (of gasterocarp): irregular in form, resembling a tuber.

unipileate (of gasterocarp development): forming one pileus.

verrucose: forming wart-like structures over the surface.
volva (in *Phallales*): remains of the peridium forming a persistant cup at base of gasterocarp.

REFERENCES

Andersson, O. (1989). Stinksvampen, *Phallus impudicus*, i Norden. *Svensk Bot. Tidskr.* 83: 219 -241.

Anonymous (1981). [colour plate of *Lycoperdon pedicellatum*]. *Z. Mykol.* 47(2): frontpage, unnumbered.

Anonymous (1992). *Naturschutz Spezial. Rote Liste der gefährdeten Grosspilze in Deutschland*. Deutsche Gesellschaft für Mykologie & Naturschutzbund Deutschland.

Arnolds, E. (1982). *Ecology and coenology of macrofungi in grasslands and moist heathlands in Drenthe, the Netherlands. 1*. Vaduz: J. Cramer (Bibl.Mycol. 83).

Arnolds, E. (1982). *Ecology and coenology of macrofungi in grasslands and moist heathlands in Drenthe, the Netherlands. 2*. Vaduz: J. Cramer (Bibl.Mycol. 90).

Askew, B. & Miller, O.K. (1977). New evidence of close relationships between *Radiigera* and *Geastrum* (Lycoperdales). *Can. J. Bot.* 55: 2693 - 2700.

Atkinson, G. F. (1911). The origin and taxonomic value of the 'veil' in *Dictyophora* and *Ithyphallus*. *Bot. Gaz.* 51: 1 - 20.

Azema, R. C. (1990). Queletia mirabilis. *Bull. Féd. Myc. Dauphiné-Savoie* 119: 20 - 22.

Badham, C.D. (1863). A Treatise on the Esculent Funguses of England. Ed. 2. London: Reeve

Barla, J.B. (1859). *Les Champignons de la Province de Nice*. Nice.

Beneke, E.S. (1963). Calvatia, calvacin and cancer. *Mycologia* 55: 257 - 270.

Berkeley, M. J. (1853). On two new genera of fungi. *Trans. Linn. Soc.* 21: 149 - 154, pl.19.

Berkeley, M.J. (1857). *Introduction to Cryptogamic Botany*. London: Bailliere.

Berkeley, M. J. (1860). *Outlines of British Fungology*. London.

Bolton, J. (1788). *An history of fungusses growing about Halifax*. Vols. I - II. Huddersfield.

Bon, M. (1987). *The Mushrooms and Toadstools of Britain and North-western Europe*. London: Hodder & Stoughton.

Brand, A. W. (1988). Profiles of Fungi 13. *Cyathus striatus* (Huds.) Willd.: Pers. *The Mycologist* 2: 110.

Breitenbach, J. & Kränzlin, F. (1986). *Fungi of Switzerland. 2. Non-gilled Fungi*. Pp412. Lucerne: Verlag Mykologia.

Brodie, H. J. (1975). *The Bird's Nest Fungi*. Pp198, 64 figs. Toronto: University of Toronto Press.

Brodie, H. J. & R. W. G. Dennis (1954). The Nidulariaceae of the West Indies. *Trans. Br. Mycol. Soc.* 37: 151 - 160, figs. 1 - 2.

Bromfield, W. A. (1843). Discovery of *Clathrus cancellatus* in Britain. *Ann. Mag. Nat. Hist.* ser.1, 12: 451.

Bronchart, R., Calonge, F.D. & Demoulin, V. (1975). Nouvelle contribution a l'étude de l'ultrastructure de la paroi sporale des Gastéromycètes. *Bull. Soc. mycol. Fr.* 91: 231 - 246.

Brøndegaard, J.V. (1983). Stinksvamp - ikke for sarte naeser og øjne. *Svampe* 8: 85 - 90.

Brøndegaard, J.V. (1987). *Folk og flora* vol. 1. Denmark: Rosenkilde & Bagger

Broome, C. E. (1870). *Scleroderma geaster* Fr., a new British fungus. *Trans. Woolhope Nat. Field Club* 1870: 252 - 253.

Bryant, C. (1782). *An Historical Account of Two Species of Lycoperdon*. London

Buczacki, S. (1989). *Fungi of Britain and Europe*. Pp 320. London: Collins New Generation Guide.

Buller, A.H. (1915). The fungus lore of the Greeks and Romans. *Trans. Brit. Mycol. Soc.* 5: 21 - 66.

Buller, A. H. R. (1933). The *Sphaerobolus* gun and its range. *Researches in Fungi* 5: 279 - 390. London: Longman, Green & Co.

Burk, W.R. (1983). Puffball usages among North American Indians. *J. Ethnobiol.* 3: 55 - 62.

Calonge, F.D. (1992). El género Bovista Pers.: Pers. (Gasteromycetes), en la península Ibérica e Islas Baleares. *Bol. Soc. Micol. Madrid* 17: 101-113.

Calonge, F. d. & Demoulin, V. (1975). Les gastéromycètes d'Espagne. *Bull. Soc. Myc. Fr.* 91: 247 - 292.

Calonge, F. D. & J. T. Palmer (1988). *Mycocalia denudata* (Fr.) Palmer, nueva para España. *Bol. Soc. Micol. Madrid* 12: 131 - 132.

Calonge, F.D. & Martín, M.P. (1990). Notes on the taxonomical delimitation in the genera Calvatia, Gastropila and Langermannia (Gasteromycetes). *Bol.Soc.Micol. Madrid* 14: 181 - 190.

Cejp, K. (1958). Nidulariales. In Pilat, A., Gasteromycetes, *Flora CSR*: 633 - 682.

Cejp, K. & J. T. Palmer (1963). The genera *Nidularia* Fr. and *Mycocalia* J. T. Palmer in Czechoslovakia and *Mycocalia sphagneti* J. T. Palmer sp. nov. from England. *Ceská Mykol.* 17: 113 - 126

Cetto, B. (1983). *Die grosse Pilzführer* 3. Trento.

Cetto, B. (1984). *Die grosse Pilzführer* 4. Trento.

Cetto, B. (1988). *Enzyklopädie der Pilze* 4. Munich: BLV Verlagsgesellschaft.

Cetto, B. (1992). *I funghi dal vero.* 7. Trento.

Chaumeton, H. (1987). *Pilze Mitteleuropas.* Gustav Fischer.

Clémençon, H., Cattin, S., Ciana, O., Morier-Genoud, R. & Scheibler, G. (1980). *Les quatre saisons des champignons* 2. Lausanne: Editions Piantanida.

Clusius, J.C. (1601). *Rariorum plantarum historia.* Antwerp.

Coccia, M., Miggliozzi, V. & C. Lavorato (1990). Studio sul genere *Scleroderma* Persoon. *Bol. d'Assoc. Micol. Ecol. Romana* 20 - 21: 3 - 59.

Coetzee, J. C. & Eicker, A. (1992). Batarrea, Battarrea, Battarraea or...???. *Mycologist* 6: 61 - 63.

Coker, W. C. & Couch, J. N. (1928). *The Gasteromycetes of the Eastern United States and Canada.* University of North Carolina Press.

Cooke, M.C. (1871). *Handbook of British Fungi.* Macmillan & Co.

Colenso, W. (1893). Bush jottings No.2 Botanical. *Trans. New Zeal. Inst.* 25: 307 - 319.

Cooke, M.C. (1862). *A plain and easy account of British Fungi.* London: Hardwicke.

Cooke, M. C. (1888). *Mutinus bambusinus* in Britain. *Grevillea* 17: 17, pl. 173.

Cunningham, G. H. (1944). *The Gasteromycetes of Australia and New Zealand.* Pp236. Dundein: John McIndoe.

Curry, F. (1864). Notes on British Fungi. *Trans. Linn. Soc., Lond.* 24: 151 - 160, pl.25.

De Candolle, A. P. (1805). In De Lamarck, J. B. & De Candolle, A. P., *Flore française* Ed. 3, vol. 2. Paris.

Demoulin, V. (1968). Gastéromycètes de Belgique: Sclerodermatales, Tulostomatales, Lycoperdales. *Bull. Jard. Bot. nat. Belg.* 38: 1 - 101.

Demoulin, V. (1969). *Les Gastéromycètes.* Bruxelles: *Les Naturalistes Belges.*

Demoulin, V. (1971). Lycoperdon norvegicum Demoulin sp.nov. A new gasteromycete with boreo-continental distribution in Europe and North America. *Norwegian Journal of Botany* 18: 161 - 167.

Demoulin, V. (1972a). Espèces nouvelles ou méconnues du genre Lycoperdon (Gastéromycètes). *Lejeunia* 62: 1 - 28.

Demoulin, V. (1972b). *Le genre Lycoperdon en Europe et en Amérique du Nord. Étude taxonomique et phytogéographique.* Doctorat en sciences botaniques, Université de Liège (distributed as xerox copy).

Demoulin, V. (1983). Clé de determination des especes d genre Lycoperdon presentes dans le sud l'Europe. *Rev. Biol.* 12: 65 - 70.

Demoulin, V. (1984). Typification of *Geastrum* Pers.: Pers. and its orthographic variant *Geaster* (Gasteromycetes). *Taxon* 33: 498 - 501.

Demoulin, V. (1993). Calvatia pachyderma (Peck) Morg. and Gastropila fragilis (Lév.) Homrich et Wright, two possible names for the same fungus. *Mycotaxon* 46: 77 - 84.

Demoulin, V. & Marriott, J.V.R. (1981). Key to the Gasteromycetes of Great Britain. *Bull. Brit. Mycol. Soc.* 15: 37 - 56.

Dennis, R.W.G. (1955). The status of *Clathrus* in England. *Kew Bull.* 10: 101 - 106.

Dennis, R.W.G., Reid, D.A. & Spooner, B.M. (1977). The fungi of the Azores. *Kew Bull.* 32: 85 -136.

Dennis, R. W. G. & Wakefield, E. M. (1946). New or interesting British fungi. *Trans. Brit. Mycol. Soc.* 29: 141 -165.

Derbsch, H. & Schmitt, J. A. (1987). Atlas der Pilze des Saarlandes. Teil 2: Nachweise, Ökologie, Vorkommen und Beschribungen. *Aus Natur und Landschaft im Saarland, sonderband* 3. Saarbrücken.

Dickinson, C. & Lucas, J. (1979). *The Encyclopaedia of Mushrooms.* Orbis.

Dörfelt, H. (1985). *Die Erdsterne. Geastraceae und Astraeaceae.* Leipzig.

Dring, D. M. (1964). British records 72. *Trans. Br. Mycol. Soc.* 47: 295 - 296.

Dring, D.M. (1966). *Tulostoma melanocyclum* Bres. British Records 93. *Trans. Brit.Mycol. Soc.* 49: 169 - 171.

Dring, D. M. (1973). Chapter 24. Gasteromycetes. In Ainsworth, G. C., Sparrow, F. K. & Sussman, A. S. (Edit.). *The Fungi. An Advanced Treatise 4B. A Taxonomic Review with Keys: Basidiomycetes and Lower Fungi.* Pp 451 - 478. London: Academic Press.

Dring, D. M. (1980). Contributions towards a rational arrangement of the Clathraceae. *Kew Bull.* 35: 1 - 96.

Dring, D. M. & Reid, D. A. (1964). British records 73. *Trans. Brit. Mycol. Soc.* 47: 296 - 297.

Dumée, P. & Maire, R. (1913). Note sur *Queletia Mirabilis* Fr. et sa découverte aux environs de Paris. *Bull. Soc. Mycol. Fr.* 29: 495 - 502.

Eckblad, F.-E. (1955). The Gasteromycetes of Norway. *Nytt Magasin for Botanikk* 4: 19 - 86.

Elborne, S. A. (1983). *Mycocalia denudata* - en ny redesvamp for Danmark. *Svampe* 7: 45.

Elborne, S. A. (1989). Danske Kiltsvampe. *Svampe* 19: 1 - 11.

Ellis, E. A. (1976). *British Fungi* Book 2. Jarrold.

Ellis, E. A. (1981). Earth-stars (Geastraceae) in Norfolk and Suffolk. *Trans. Norfolk & Norwich nat. Soc.* 25: 145 - 152.

Ellis, M. B. & Ellis, J. P. (1990). *Fungi without Gills (Hymenomycetes and Gasteromycetes). An Identification Handbook.* Pp266, 49 pl. London: Chapman and Hall.

Emboden, W.A. (1974). *Bizarre Plants. Magical, Monstrous, Mythical.* London: Macmillan.

Favre, J. (1937). Champignons rares ou peu connus des Hauts-Marais Jurassiens. *Bull. Soc. Myc. France* 53: 271-296.

Findlay, W.P.K. (1982). *Fungi. Foklore, Fiction and Fact.* Richmond Publishing Co.

Fischer, E. (1887). Zur entwicklungsgeschichte der Fruchkörper einiger Phalloideen. *Ann. Jard. Bot. Buitenz.* 6: 1 - 51, pl. 1 - 5.

Fischer, E. (1900). Plectobasidiineae. In Engler, A. & Prantl, K.*Nat. Pflanzenfam.* 1, Abt.1 Lief. 193.

British Puffballs, Earthstars and Stinkhorns

Folkard, R. (1884). *Plant Lore, Legends, and Lyrics*. London.

Francis, S. M. (1990). Spores and splashes. *Mycologist* 4: 41 - 42.

Friend, H. (1886). *Flowers and flower lore*. Ed. 3. London.

Fuhrer, B. (1985). *A Field Companion to Australian Fungi*. Pp 162. Hawthorn, Victoria: Five Mile Press.

Gerard, J. (1597). *The Herball or general historie of plantes*. London: John Norton.

Gerard, J. & Johnson, T. (1633). *The Herball or generall historie of plantes*. London.

Gerhardt, E. (1985). *Pilze. Band 2: Röhrlinge, Porlinge, Bauchpilze, Schlauchpilze und andere*. Pp320. München, Wien, Zürich: BLV Verlagsgesellschaft.

Gibbs, T. (1908). *Bovistella paludosa* Lév. A puffball new to Britain. *Naturalist, Hull* 1908: 457.

Greville, R. K. (1823). *Scottish Cryptogamic Flora or coloured figures and descriptions of cryptogamic plants found in Scotland and belonging chiefly to the order Fungi and intended to serve as a continuation of English Botany* 1: pl.1 - 60. Edinburgh: Maclachnan.

Guzman, G. (1970). Monografia del genero *Scleroderma* Pers. emend. Fr. *Darwiniana* 16: 233 - 407.

Handke, H. H. (1963). *Dictyophora duplicata* (Bosc) E. Fischer. *Mykol. Mitteil.* 7: 33 - 44.

Hansen, G. R. (1986). Aertetrøffel - et sjaeldent (og) grimt syn. *Svampe* 13: 25 - 29, figs. 1 - 4.

Hennebert, G. L. (1973). *Botrytis* and *Botrytis* - like genera. *Persoonia* 7: 183 - 204.

Henriksen, N.T. (1976). Lycoperdonosis. *Acta paediatr. Scand.* 65: 643 - 645.

Herrera, T. (1953). Un Hongo Nuevo Procedente del estado de San Luis Potosi. Battarreoides potosinus gen. nov. sp. nov. *An. Inst. Biol. Méx.* 24: 41 - 46.

Herrman, M. (1962). Der Tintenfischpilz, *Anthurus archeri* (Berk.) E. Fischer, erstmals in der DDR beobachtet. *Mykol. Mitteilungsbl.* 6: 4 - 9.

Hooker, J.D. (1854). *Himalayan Journals*. vol. 2. London: John Murray.

Hudson, G. (1878). *Flora Anglica; exhibens plantas per regnum Britanniae sponte crescentes*. Edit.2, 2: 335 - 690. London.

Ing, B. (1985). *Pisolithus* in Ireland. *Bull. Brit. Mycol. Soc.* 19: 57 - 58.

Ing, B. (1989). First steps. Puffballs. *Mycologist* 3: 126 - 127.

Ing, B. (1993). Towards a red list of endangered European macrofungi. In Pegler & al. (eds.). Fungi of Europe: Investigation, recording and conservation 231 - 237. Royal Botanic Gardens, Kew.

Ingold, C. T. (1972). *Sphaerobolus*: the story of a fungus. *Trans. Brit. Mycol. Soc.* 58: 179 - 195.

Jahn, H. (1972). Ein Massenvorkommen von *Nidularia farcta* im östlichen Westfalen. *Westf. Pilzbr.* 9: 16 - 19, figs. 1 - 4.

Jahn, H. (1979). *Pilze die an Holz wachsen*. Pp 268, 222 pl. Bussesche Verlagshandlung.

James, R. (1747). Pharmocopeia Universalis.

Jay, E., Noble, M. & Hobbs, A.S. (1992). *A Victorian Naturalist. Beatrix Potter's Drawings from the Armitt Collection*. London: Warne.

Jeppson, M. (1984). *Släktet Lycoperdon i Sverige*. Sveriges Mykologiska Förening, Pp.47.

Jeppson, M. (1987). Notes on some Spanish gasteromycetes. *Bol. Soc. Micol. Madrid* 11(2): 267-282.

Jeppson, M. & Demoulin, V. (1989). Lycoperdon atropurpureum found in Sweden. *Opera Bot.* 100: 131 - 134.

Jülich, W. (1981). Higher Taxa of Basidiomycetes. *Bibl. Mycol.* 85: 1 - 485

Jülich, W. (1983). Sporenstruktur bei Gastromycetes. 2. Queletia. *Int. J. Myc. Lich.* 1: 169 - 174.

Jülich, W. (1984). *Die Nichtblätterpilze, Gallertpilze und Bauchpilze.* In Gams, H. (ed.), *Kleine Krypogamenflora* 2b, 1. Stuttgart: Gustav Fischer.

Kers, L. E. (1978). Tulostoma niveum sp. nov. (Gasteromycetes), described from Sweden. *Bot. Notiser* 131: 411 - 417.

Knudsen, H. (1986). *Mutinus ravenelii* - en ny dansk stinksvamp. *Svampe* 14: 64 - 65, figs. 1 - 2.

Kobler, B. (1982). Lycoperdon echinatum Pers., Igelstäubling. *Schweiz. Z. Pilzk.* 60: 222 & plate.

Kobler, B. (1986). Lycoperdon mammaeforme Pers., Flockenstäubling. *Schweiz. Zeitschr. Pilzk.* 86: 30 & plate.

Kosonen, L. (1986). Mutinus ravenelii (Gasteromycetes) in Finland. *Lutukka* 2: 15 - 19.

Kotlaba, F. (1955). Prásivka uherská - Bovista hungarica Holl. - houba nasich polí. *Ceská Mykol.* 9: 169 - 171.

Kreisel, H. (1962). Die Lycoperdaceae der Deutschen Demokratischen Republik. *Feddes Repertorium* 64: 89 - 201, pl. 1 - 9.

Kreisel, H. (1967). Taxonomisch-pflanzengeographische Monographie der Gattung Bovista. *Beih. z. Nova Hedwigia* 25: 1-244, pl. 1-70.

Kreisel, H. (1969). *Grundzüge eines natürlichen System der Pilze.* Pp 245, 8 pl. Jena: Gustav Fischer Verlag.

Kreisel, H. (1973). Die Lycoperdaceae der DDR. *Bibl. Mycol.* 36.

Kreisel, H. (1987). *Pilzflora der Deutschen Demokratischen Republik. Basidiomycetes (Gallert-, Hut- und Bauchpilze).* Gustav Fischer.

Kreisel, H. (1989). Studies in the Calvatia complex (Basidiomycetes). *Nova Hedvigia* 48: 281 - 296.

Kreisel, H. (1992). An emendation and preliminary survey of the genus Calvatia (Gasteromycetidae). *Persoonia* 14(4): 431 - 439.

Kreisel, H. (1993). A key to *Vascellum* (Gasteromycetidae) with some floristic notes. *Blyttia* 51: 125 - 129.

Kreisel, H. & Calonge, F.D. (1993). *Calvatiella* Chow, a synonym for *Bovistella* Morgan. *Mycotaxon* 48: 13 - 25.

Krieglsteiner, G. J. (1992). Das neue europäische Areal des Tintenfischpilzes - Clathrus archeri (Berk.) Dring. Beit. z. Kenntn. der Pilze Mitteleuropas 8: 29 - 64.

Kubicka, J. (1984). Zum *Battarraea* - Fund in der CSSR. *Mykol. Mitteilungsbl.* 27: 54.

Lange, M. (1990). Arctic Gasteromycetes II. Calvatia in Greenland, Svalbard and Iceland. *Nord. J. Bot.* 9: 525 - 546.

Lange, M. & Hora, F. B. (1965). *Mushrooms & Toadstools.* London: Collins.

Lefevre, A. (1982). Battarea phalloides Persoon. *Bull. Féd. Mycol. Dauphiné-Savoie* 22: 7 - 9.

Leightley, L. E. & Eaton, R. A. (1979). *Nia vibrissa* - a marine white rot fungus. *Trans. Brit. Mycol. Soc.* 73: 35 - 40.

Léveillé, J. H. (1848). Fragments mycologique. *Ann. Sci. Nat., Bot.* sér. 3, 9: 119 - 144.

Lincoff, G. & Mitchell, D.H. (1977). *Toxic and Hallucinogenic Mushroom Poisoning. A handbook for physicians and mushroom hunters.* Van Nostrand Reinholt.

Lloyd, C.G. (1910). A Bovistella with a Geaster mouth. *Mycol. Writings* 3: 452.

Lohwag, H. (1933). Mykologische Studien VIII. Bovista echinella Pat. und Lycoperdon velatum Vitt. - *Beih. Bot. Zentralbl.* 51, Abt.1: 269-286.

MacMillan, H. (1861). *Footnotes from the page of nature or first forms of vegetation.* Cambridge & London: MacMillan.

Maire, R. (1930). Etudes mycologiques. *Bull. Soc. Mycol. Fr.* 46: 215 - 244, pl.X.

Marchand, A. (1973). *Champignons du Nord et du Midi.* Pp273, pl.101 - 200. Perpignan, France: Soc. Mycol. Pyren. Medit.

Marchand, A. (1976). *Champignons du Nord et du Midi 4.* Pp261, pl.301 - 400. Perpignan, France: Soc. Mycol. Pyren. Medit.

Martín Estaban, P. (1988). *Aportacion al conocimiento de las Higroforaceas y los gasteromicetes de Cataluña.* Sociedad Catalana de Micologia, Universitat de Barcelona.

Martin, G.W. (1936). A Key to the Families of Fungi. *Stud. nat. Hist. Iowa Univ.* 17: 83 - 115.

Massee, G. (1889). A monograph of the British Gasteromycetes. *Ann. Bot.* 4: 1 - 103, pl. 1 - 4.

Massee, G. (1897). Mycologic flora of the Royal gardens, Kew. *Bull. Misc. Inf. Kew* 1897: 115 - 167.

Maublanc, A. & Malençon, G. (1930). Recherches sur le Battarraea Guicciardiniana Ces. *Bull. soc. Mycol. Fr.* 46: 43 - 73.

Michael, E., Hennig, B. & Kreisel, B. (1986). *Handbuch für Pilzfreunde* vol. 2. Stuttgart, Fischer.

Miller, W. (1884). *A Dictionary of English Names of Plants.* J. Murray.

Miller, O.K. & Miller, H.H. (1988). *Gasteromycetes. Morphological and Developmental Features with keys to Orders, Families, and Genera.* Mad River Press.

Monthoux, O. (1982). Micromorphologie des spores et capillitiums des Gastéromycètes des stations xériques de la region de Genève, étudiée au microscope électronique à balayage (SEM).*Candolleana* 37: 63- 99.

Monthoux, O. & Röllin, O. (1984). La flore fongique des stations xeriques de la region de Geneve V. Lycoperdaceae: Genres Bovista (fin), Lycoperdon, Vascellum et Geastraceae: genre Geastrum (Basidiomycotina, Gasteromycetes). *Mycologia Helvetica* 1(3): 189 - 208.

Moravec, Z. (1958a). Queletia Fr. In Pilát, A. (ed.) Gasteromycetes. *Flora CSR* B 1: 615 - 617.

Moravec, Z. (1958b). Battarrea Pers. In Pilát, A. (ed.). Gasteromycetes. *Flora CSR* B 1: 621 - 624.

Mornand, J. (1989). Les Gasteromycetes de France (5 - Tulostomatales). *Doc. mycol.* 19, fasc. 76: 1 - 18.

Moser, M. & Jülich, W. (1988). *Colour Atlas of Basidiomycetes.* Lief. 6. Gustav Fischer.

Moser, M. & Jülich, W. (1989). *Colour Atlas of Basidiomycetes.* Lief. 7. Gustav Fischer.

Murrill, W. A. (1912). Illustrations of Fungi XI. *Mycologia* 4: 163 - 169, pl.68.

Oberwinkler, F. & Bauer, R. (1989). The systematics of gasteroid, auricularioid Heterobasidiomycetes. *Sydowia* 41: 224 - 256.

Oberwinkler, F. Bauer, R. & Bandoni, R. J. (1990). Heterogastridiales: a new order of Basidiomycetes. *Mycologia* 82: 48 - 58.

Ortega, A. & Buendia, A.G. (1989). Estudio del complejo Bovista aestivalis (Bon.) Demoulin - B. pusilla (Batsch) Pers. sensu Kreisel en España. *Cryptogamie, Mycol.* 10: 9-18.

Ortega, A., Buendia, A. G. & Calonge, F. D. (1985): Estudio de algunas especies interessantes del genero *Lycoperdon* (Gasteromycetes) en España. *Bol. Soc. Micol. Castellana* 9: 141 - 148.

Oso, B.A. (1976). Phallus aurantiacus from Nigeria. *Mycologia* 68: 1076 - 1082.

Ott, J., Guzmán, G., Romano, J. & Díaz, J.L. (1975). Nuevos datas sobre los supuestos Licoperdáceos sicotrópicos y dos casos intoxicación provocado por hongos del género Scleroderma en México. *Bol. Soc. Mex. Mic.* 9: 67 - 76.

Pacioni, G. (1983). Battarraea phalloides (Dicks.) Pers. *Boll. Gr. mic. Bres.* 26: 93 - 96.

Palmer, J. T. (1952). Lancashire and Cheshire Earth-stars. *Proc. Liverp. Nat. Field Cl.* 1952: 28 - 31.

Palmer, J. T. (1955). Observations on Gasteromycetes 1 - 3. *Trans. Brit. Mycol. Soc.* 38: 317 - 334.

Palmer, J. T. (1958). Observations on gasteromycetes. VI. Three British species of *Nidularia* Fr. section *Sorosia* Tul.: ecology and distribution. *Trans. Brit. Mycol. Soc.* 41: 55 - 63.

Palmer, J. T. (1963). Deutsche und andere Arten der Gattung *Mycocalia*. *Zeitschr. Pilzk.* 28: 15 - 21, figs 1-10.

Palmer, J.T. (1968). A Chronological Catalogue of the Literature to the Briish Gasteromycetes. *Nova Hedwigia* 15: 65 - 178.

Park, D. (1981). *Lysurus gardneri*, a new Irish record. *Bull. Brit. Mycol. Soc.* 15: 142 - 143.

Pegler, D. N. (1990).*Field Guide to the Mushrooms & Toadstools of Britain & Europe*. Pp192. London: Kingfisher Books.

Pegler, D.N. (1993). False Truffles (Basidiomycotina). In Pegler, D. N., Spooner, B. M. & Young, T. W. K., *British Truffles. A Revision of British Hypogeous Fungi*. Pp216, 26pl., 30 figs. Kew.

Pegler, D. N., Spooner, B. M. & Young, T. W. S. (1993). *British Truffles. A revision of British hypogeous fungi*. Pp. 216. Kew: Royal Botanic Gardens.

Perreau, J. (1986). 'Battaraea phalloides'. *Bull. Soc. mycol. Fr.* 102: Atlas, pl. 245.

Petch, T. (1908). The Phalloideae of Ceylon. *Ann. Roy. Bot. Gard. Peradeniya* 4: 139 - 184.

Phillips, R. (1981). *Mushrooms and other fungi of Great Britain & Europe*. Pp 288. London: Pan Books Ltd.

Philipps, W. & Plowright, C.B. (1876). New and rare British Fungi. *Grevillea* 4: 118 - 124.

Pilat, A. (1937). Contribution à la connaissance des Basidiomycètes de la péninsule des Balkans. *Bull. Soc. Mycol. Fr.* 53: 81 - 104.

Pilat, A. (1958). Gasteromycetes. *Flora CSR* B-1: Pp862. Praha.

Plowright, C. B. (1881). *Geaster coliformis* (Dickson). *Trans. Norfolk Norwich Nat. Soc.* 3: 266 - 267.

Poelt, J. & Jahn, H. (1963). *Mitteleuropäische Pilze*. Hamburg: J.Cramer

Ponce de Lyon, P. (1968). A revision of the family Geasteraceae. *Fieldiana, Bot.* 31: 301 - 349.

Porter, D. & Farnham, W. F. (1986). *Mycaureola dilseae*, a marine basidiomycete parasite of the red alga, *Dilsea carnosa*. *Trans. Brit. Mycol. Soc.* 87: 575 - 582.

Pouzar, Z. (1958). Tulostoma. In Pilát, A (ed.), Gasteromycetes. *Flora CSR* B 1: 589 - 613.

Rai, B.K., Ayachi, S.S. & Rai, A. (1993). A note on ethno-myco-medicines from Central India. *Mycologist* 7: 192 - 193.

Ramsbottom, J. (1916a). Battarrea phalloides in Britain. *J. Bot., Lond.* 54: 105 - 107.

Ramsbottom, J. (1916b). Battarrea phalloides Pers. in Britain. *J. Bot., Lond.* 54: 198 - 199.

Ramsbottom, J. (1953). *Mushrooms & Toadstools. A Study of the Activities of Fungi*. London: Collins New Naturalist Series.

Rauschert, S. (1986). Proposal to conserve the spelling *Battarraea* (Fungi: Gasteromycetes). *Taxon* 35: 733 - 736.

Ray, J. (1724). *Synopsis methodica stirpium Britannicarum* Ed. 3. London.

Rea, C. (1904). Notes on two Phalloideae new to Europe. *Trans. Brit. Mycol. Soc.* 2: 57 - 59, pl.3.

Rea, C. (1909). New and rare British fungi. *Trans. Brit. Mycol. Soc.* 3: 124 - 130.

Rea, C. (1912). *Glischroderma cinctum* Fckl. *Trans. Brit. Mycol. Soc.* 4: 64 - 65, pl.2.

Rea, C. (1912). British Geasters. *Trans. Brit. mycol. Soc.* 3: 351 - 353.

Rea, C. (1922). *British Basidiomycetae. A Handbook to the Larger British Fungi* Pp799. Cambridge University Press.

Rea, P. M. (1942). Fungi of Southern California. I. *Mycologia* 34: 563 - 574.

Reid, D. A. & Dring, D. M. (1964). British records. 71. *Ileodictyon cibarius. Trans. Brit. Mycol. Soc.* 47: 293 - 295.

Reid, D.A. (1985). The status of *Ileodictyon cibarius* in Britain. *Bull. Brit. Mycol. Soc.* 19: 126

Richter, W. & Müller, G. K. (1983). Der Stelzen-Stäubling - Batttarraea phalloides - neu für die DDR. *Mykol. Mitteilungsblatt* 26: 61 - 63.

Rinaldi, A. & Tyndalo, V. (1972). *Mushrooms and other fungi.* Pp 333. London: Hamlyn.

Rolfe, R.T. & Rolfe, F.W. (1925). *The Romance of the Fungus World.* London: Chapman & Hall.

Rotheroe, M., Hedger, J. & J. Savidge (1987). The fungi of Ynyslas sand-dunes. *Mycologist* 1: 15 - 17.

Runge, A. & Gröger, F. (1990). Neue Funde von Bovista limosa in Deutschland. *Myk. Mitt. bl.* 33: 91 - 94.

Ryman, S. & Holmåsen, I. (1984). *Svampar. Enfälthandbok.* Stockholm: Interpublishing.

Sarasini, M. (1991). Appunti sul genere *Scleroderma* Pers. *Riv. Micol.* 34: 34: 119 - 130.

Sarasini, M. (1992). Appunti sull' Ordine Phallales, 1 Partie: Caratteri generali dell' ordine: Famiglia Phallaceae. *Riv. Micol.* 35: 33 - 42, pl.1 - 9.

Schmidel, D. C. C. (1793). *Icones Plantarum et Analyses Partium.* Ed.2. Erlangae.

Schmitt, J. A. (1978). Zur Verbreitung und Ökologie epigäischer Gasteromycetes (Bauchpilze) im Saarland. *Abh. Arb. Gem. tier-u. pfl. geogr. Heimatforsch. Saarl.* 8: 13 - 60.

Sebek, S. (1958). Sclerodermataceae. In Pilat, A, Gasteromycetes. *Flora CSR* B-1: Pp 558 -573. Praha.

Smith, A.H. (1951). *Puffballs and their allies in Michigan.* Ann Arbor: University of Michigan Press.

Smith, A. H. (1974). The genus Vascellum (Lycoperdaceae) in the United States. *Bull. Soc. Linn. Lyon* 43, num. spe.: 407 - 419.

Smith, H.H. (1933). Ethnobotany of the forest Potawatomi indians. *Bull. Publ. Mus. City of Milwaukee* 7: 1 - 230.

Smith, J. (1882). *Dictionary of Popular Names of Economic Plants.* MacMillan & Co.

Sowerby, J. (1796). *Coloured Figures of English Fungi* 1: pl. 1 - 120. London: J. Davis.

Sowerby, J. (1801). *Coloured Figures of English Fungi* 3: pl. 241 - 400. London: R. Wilks.

Sowerby, J. (1814). *Coloured Figures of English Fungi* Suppl.: pl. 401 - 440.

Stanek, V. J. (1958). Geastraceae. In Pilat, A. (1958). Gasteromycetes. *Flora CSR* B-1: Pp 392 - 526. Praha.

Steele, A.B. (1888). Fungus folk-lore. *Trans. Edinbugh Field Nats Microscop. Soc.* 2 (2): 175 - 183.

Stevens, E.L. & Kidd, M.M. (1953). *Some South African Poisonous and Edible Fungi.* Longmans, Green & Co.

Sunhede, S. (1989). Geastraceae (Basidiomycotina).Morphology, ecology, and systematics with special emphasis on the North European species. *Synops. Fung.* 1. Oslo: Fungiflora.

Swanton, E.W. (1917). Economic and folk lore notes. *Trans. Brit. Mycol. Soc.* 5: 408 - 409.

Tulasne, L. R. & C. Tulasne(1844). Recherches sur l'organisation et la mode de fructification des champ[ignons de la tribu des Nidulariées, suives d'un Essai Monographiques. *Ann. Sci., Nat., Bot.* sér. 3, 1: 41 - 107, pl. 3 - 8.

Ulbrich, E. (1943). Mutinus ravenelii (Berk. & Curt.) E. Fischer, eine für Europa neue Phallaceae. *Notizblatt Bot. Grad. Mus. Berlin-Dahlem* 15: 820 - 824.

Umezawa, H.T. et al. (1975). A new antibiotic, calvatic acid. *J. antibiot.* 28: 87 - 90.

Vaillant, S. (1727). *Botanicon Parisiense.* Pp 205, pl. 33. Leiden & Amsterdam: Verbeek & Lakemann.

Van den Eynden, V.; Vernemmen, P. & Van Damme, P. (1992). *The Ethnobotany of the Topnaar.* Universiteit Gent.

Vesterholt, J. & Sørensen, K. (1989). Blaekspruttesvamp (*Clathrus archeri*) fundet i Danmark. *Svampe* 19: 16 - 17.

Wakefield, E. M. (1918). New and rare British fungi. *Bull. Miss. Inf. Kew* 1918: 229 - 233.

Wakefield, E. M. & Dennis, R. W. G. (1981). *Common British Fungi. A guide to the more common larger Basidiomycetes of the British Isles.* Hindhead: Saga Publishing Co. Ltd.

Wasson, R.G., Hofman, A. & Ruck, C.A.P. (1978). The Road to Eleusis. Ethnomycological Studies 4. Wolff.

Wasson, V.P. & Wasson, R.G. (1957). *Mushrooms, Russia and History.* New York, Pantheon Books

Watling, R. (1975). Prehistoric Puff-balls. *Bull. Br. mycol. Soc.* 9: 112 - 114.

Watling, R. & Seaward, M. R. D. (1976). Some Observations on Puff-balls from British Archaeological Sites. *J. Archaeol.* Sci. 3: 165 - 172.

Weber, N.S. & Smith, A.H. (1985). *A field guide to southern mushrooms.* Ann Arbor: University of Michigan Press.

White, V. S. (1901). The Tylostomaceae of North America. *Bull. Torrey. Club* 28: 421 - 444.

Withering, W. (1792). *A Botanical Arrangement of all the Vegetables naturally growing in Great Britain.* Ed. 2, vol. 3. Birmingham.

Woodward, T. J. (1784). An account of a new plant, of the order of Fungi. *Phil. Trans. Roy. Soc. London* 74: 423 - 427.

Wright, J. E. (1987). The Genus *Tulostoma* (Gasteromycetes) - A World Monograph. *Bibl. Mycol.* 113: 1 - 338.

Wright, J. E. (1989). South American Gasteromycetes. III. The Genus *Queletia* Fr. *Crypt. Bot.* 1: 26 - 31.

Ying, J., Mao, X., Ma, Q., Zong, Y. & Wen, H. (1987). *Icones of Medicinal Fungi from China.* Beijing, Science Press.

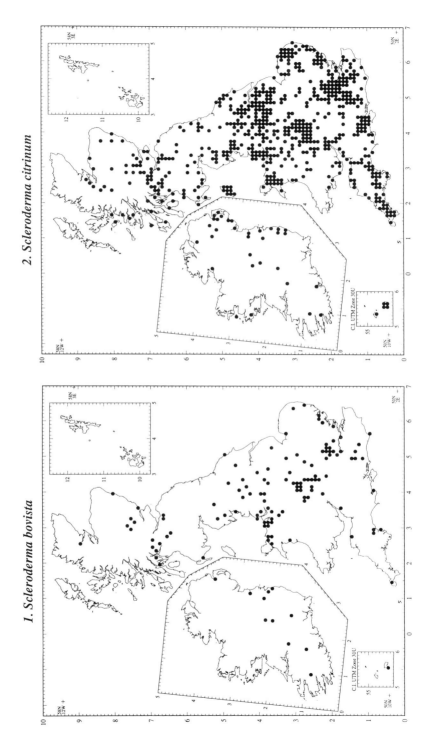

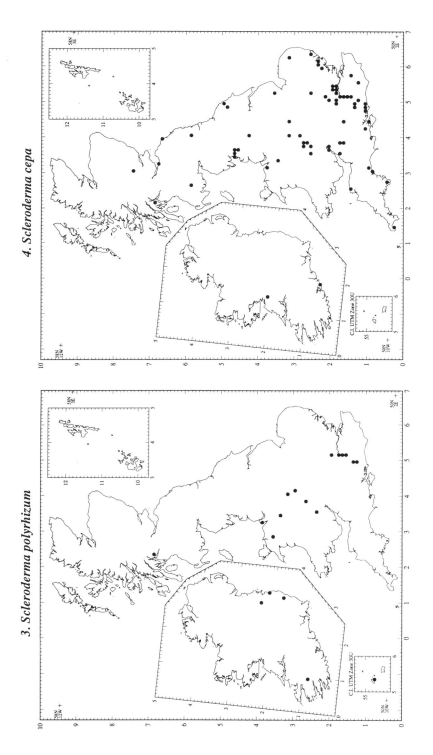

British Puffballs, Earthstars and Stinkhorns

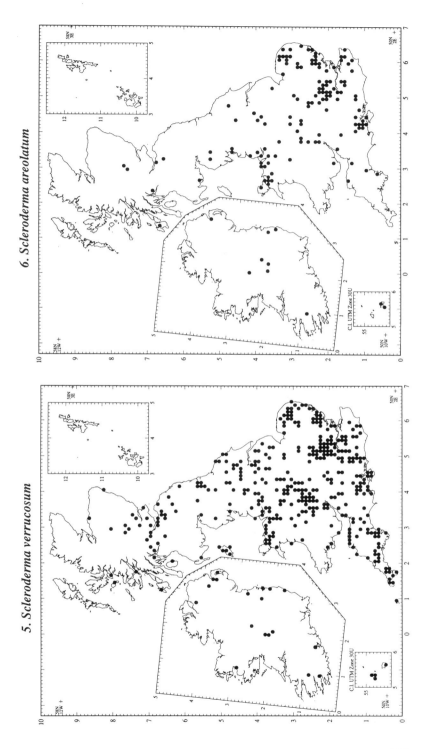

5. Scleroderma verrucosum

6. Scleroderma areolatum

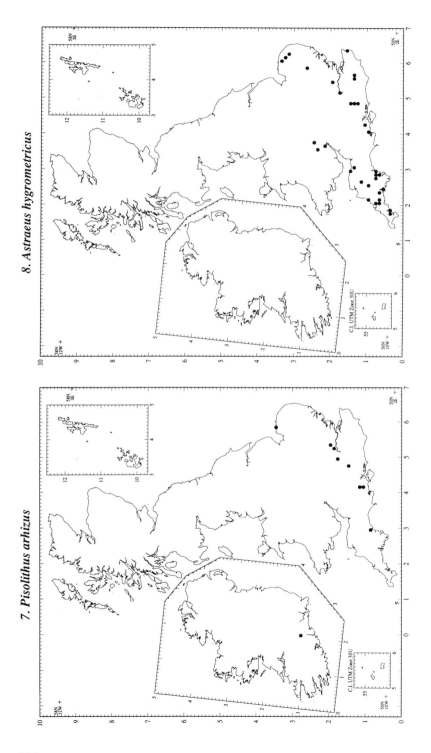

British Puffballs, Earthstars and Stinkhorns

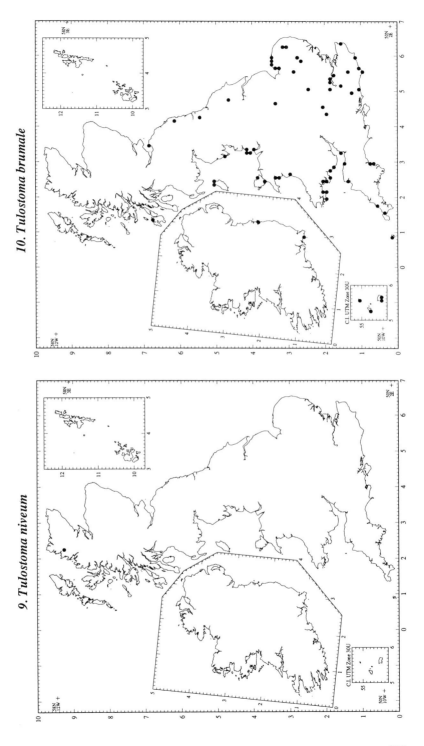

9. Tulostoma niveum

10. Tulostoma brumale

British Puffballs, Earthstars and Stinkhorns

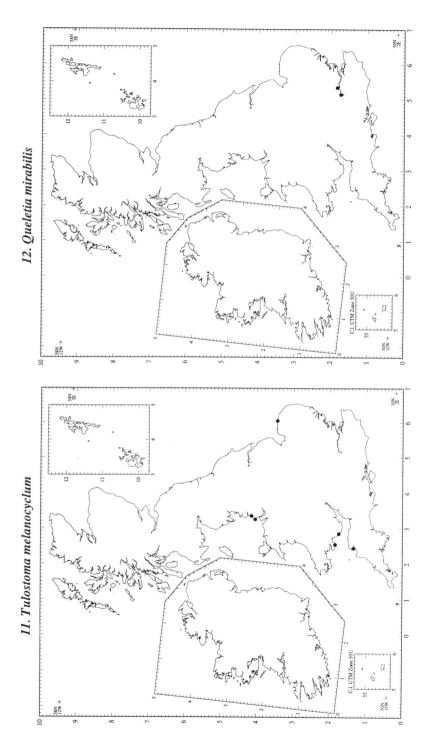

12. Queletia mirabilis

11. Tulostoma melanocyclum

British Puffballs, Earthstars and Stinkhorns

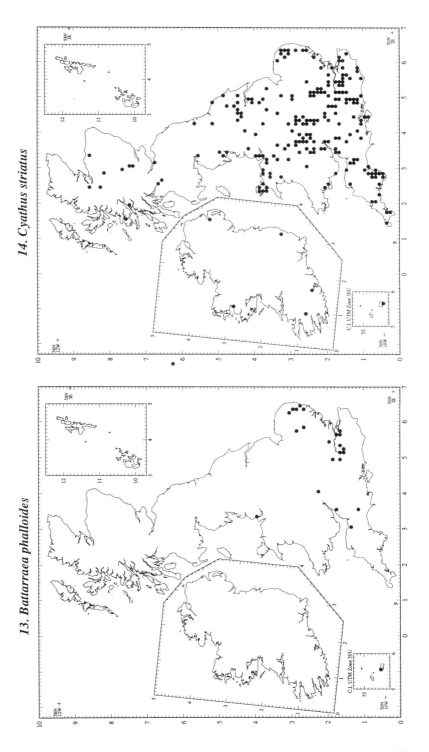

14. *Cyathus striatus*

13. *Battarraea phalloides*

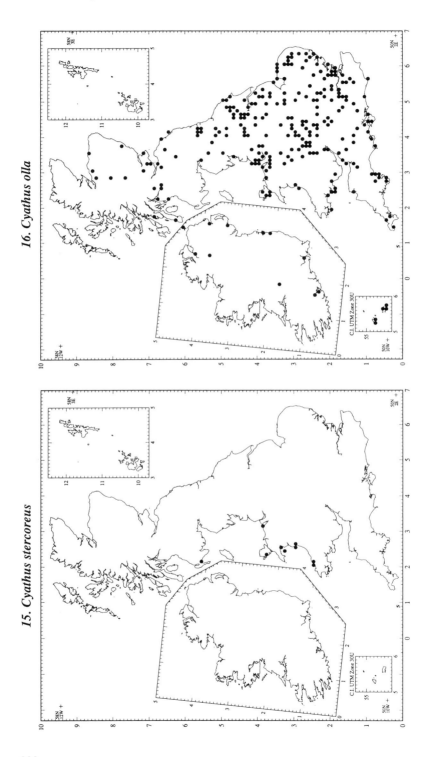

15. Cyathus stercoreus

16. Cyathus olla

British Puffballs, Earthstars and Stinkhorns

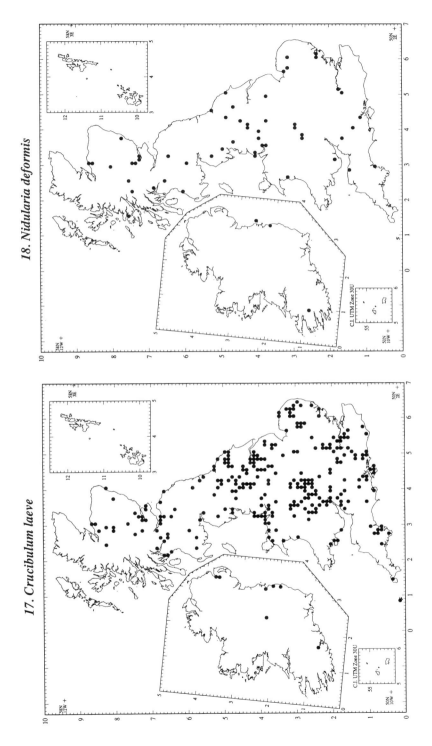

18. *Nidularia deformis*

17. *Crucibulum laeve*

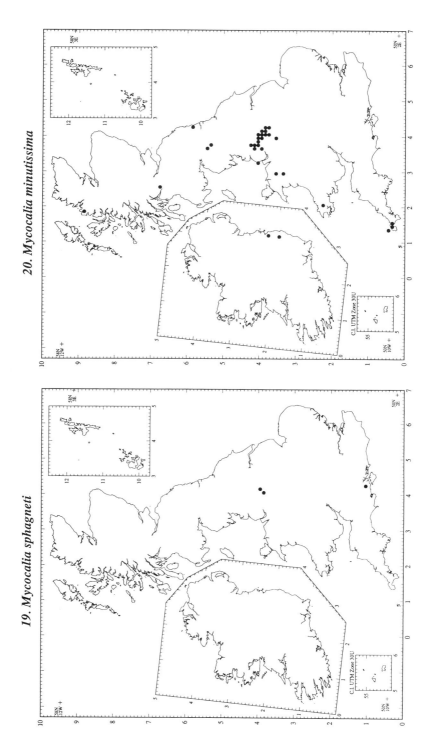

British Puffballs, Earthstars and Stinkhorns

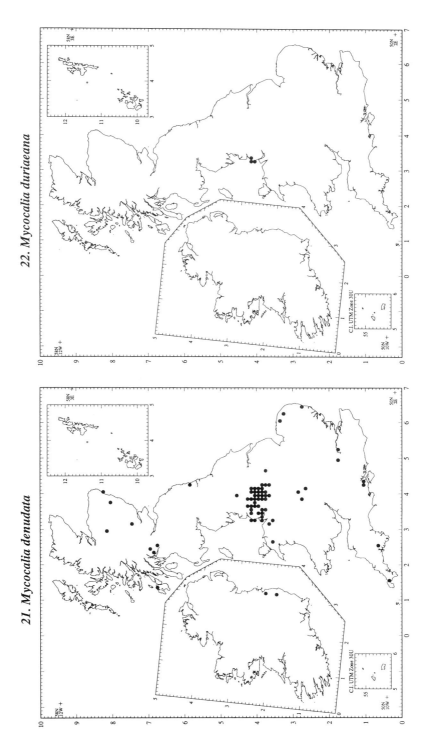

21. Mycocalia denudata

22. Mycocalia duriaeana

British Puffballs, Earthstars and Stinkhorns

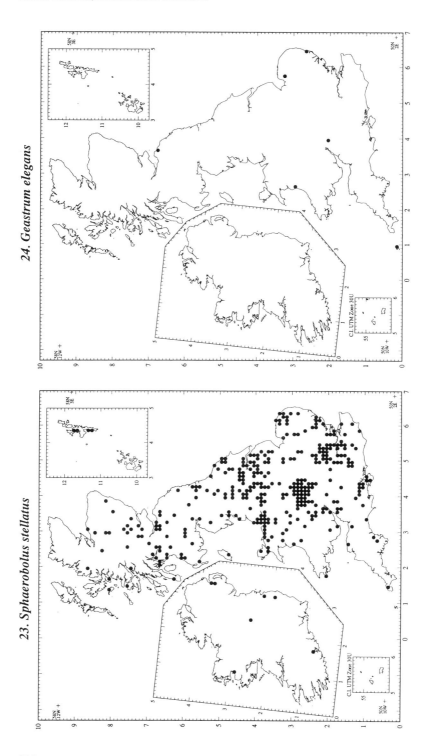

24. *Geastrum elegans*

23. *Sphaerobolus stellatus*

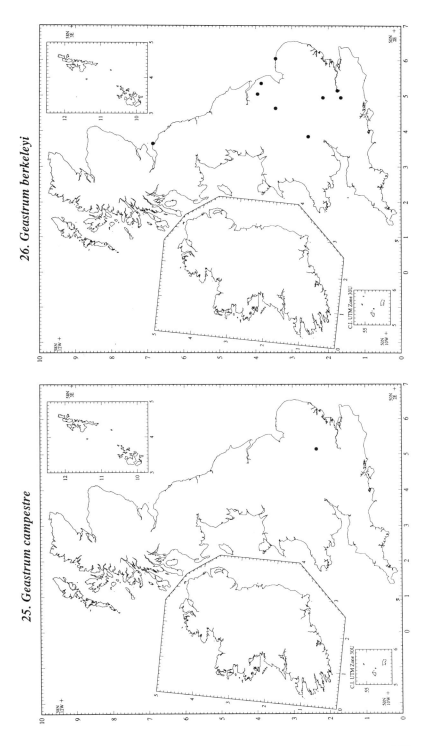

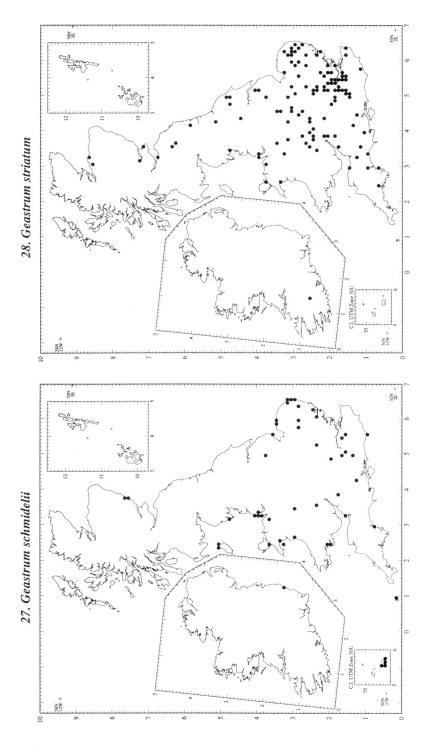

27. *Geastrum schmidelii*

28. *Geastrum striatum*

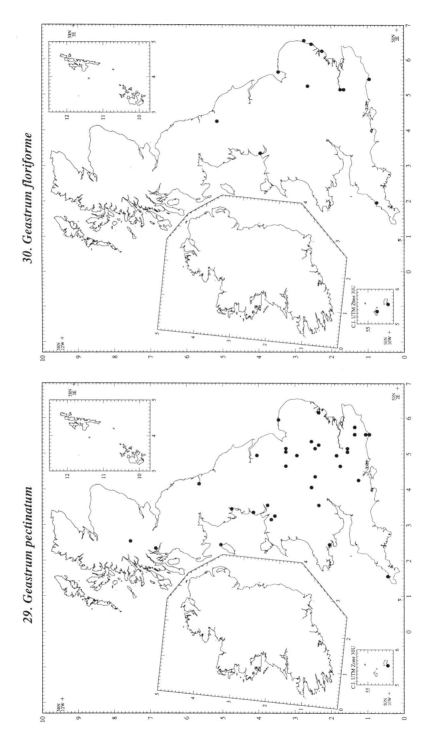

29. Geastrum pectinatum

30. Geastrum floriforme

British Puffballs, Earthstars and Stinkhorns

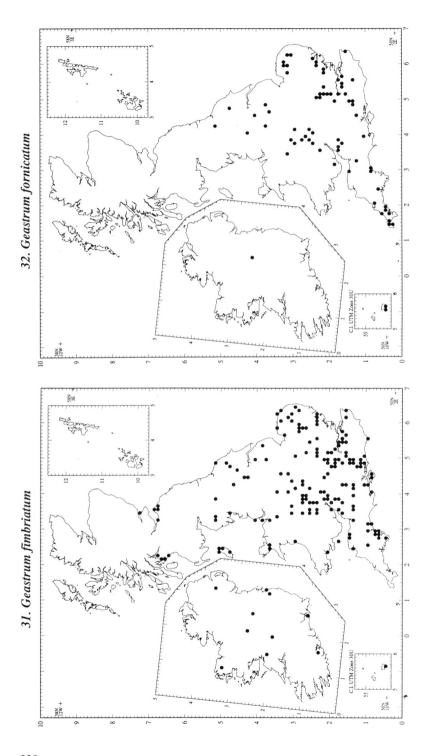

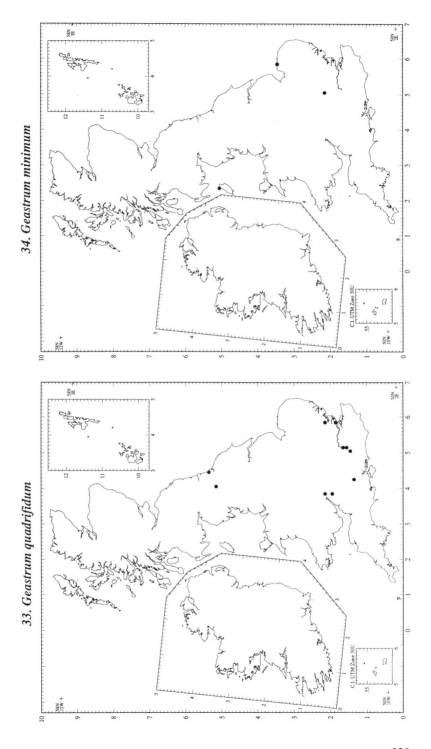

34. Geastrum minimum

33. Geastrum quadrifidum

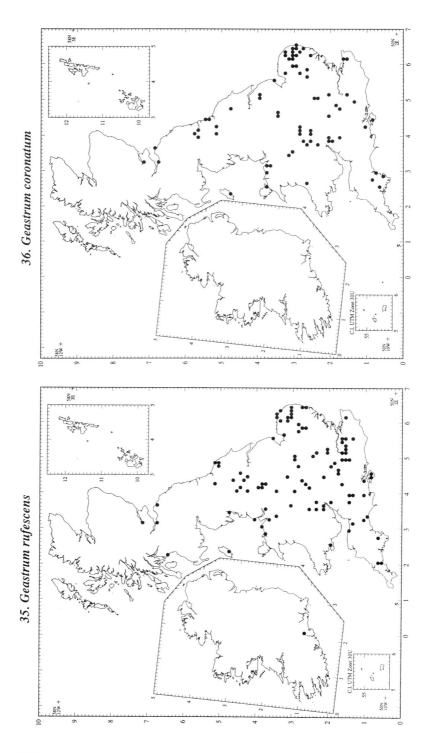

35. *Geastrum rufescens*

36. *Geastrum coronatum*

British Puffballs, Earthstars and Stinkhorns

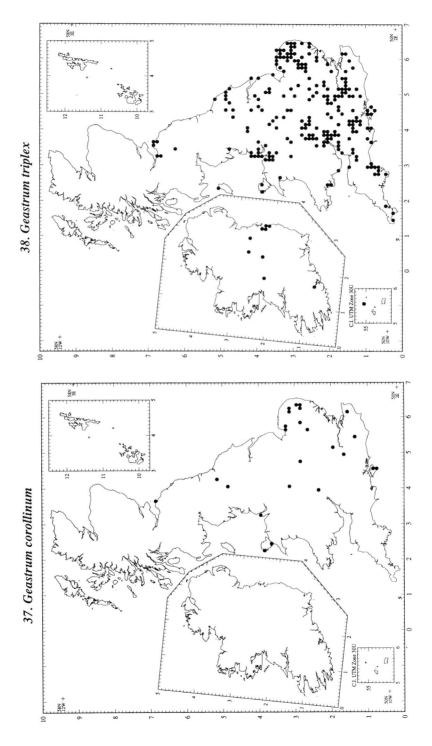

38. *Geastrum triplex*

37. *Geastrum corollinum*

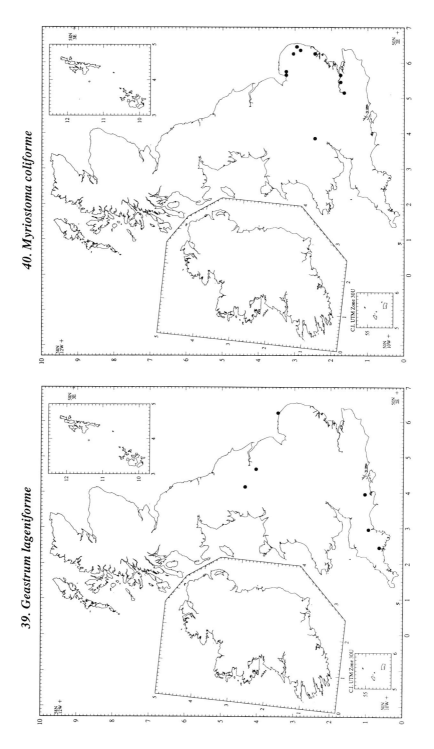

British Puffballs, Earthstars and Stinkhorns

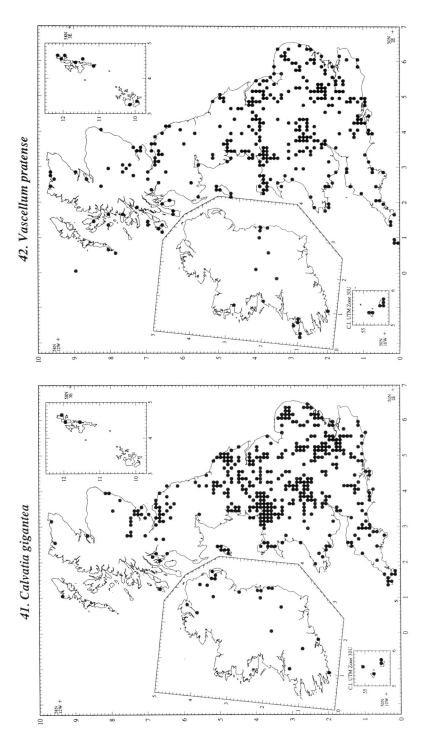

41. Calvatia gigantea

42. Vascellum pratense

British Puffballs, Earthstars and Stinkhorns

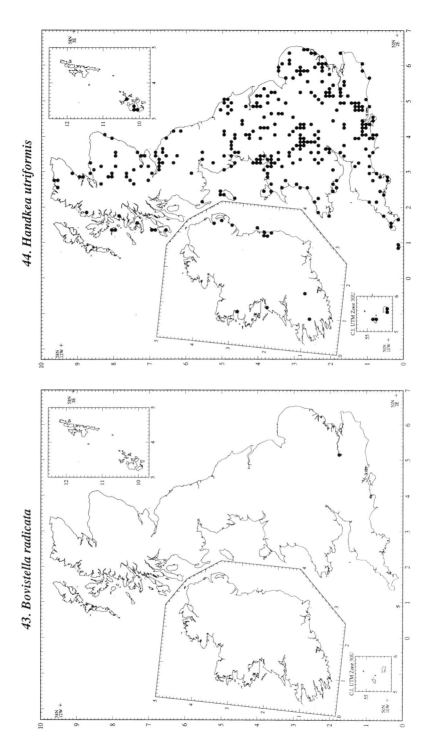

British Puffballs, Earthstars and Stinkhorns

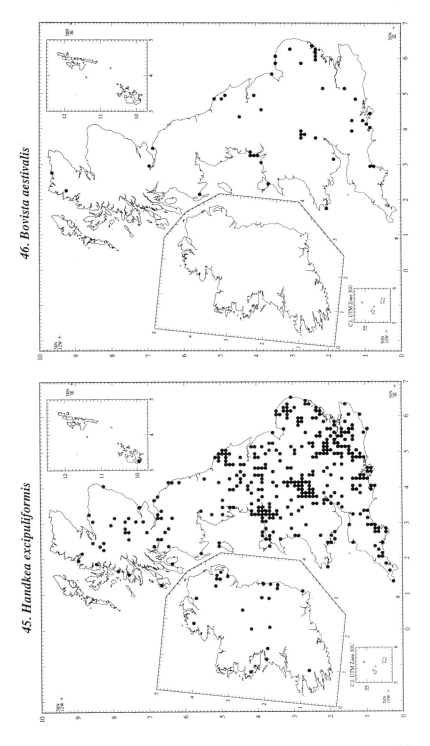

46. Bovista aestivalis

45. Handkea excipuliformis

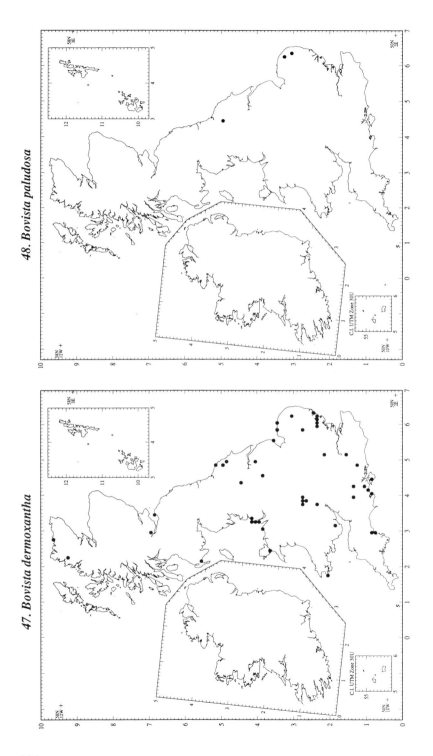

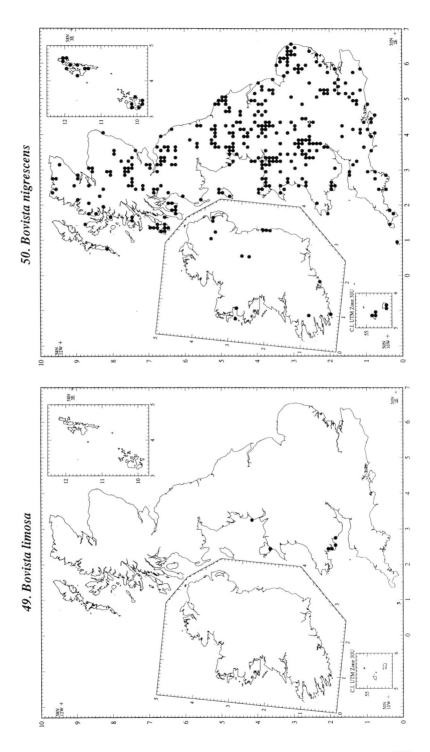

50. *Bovista nigrescens*

49. *Bovista limosa*

British Puffballs, Earthstars and Stinkhorns

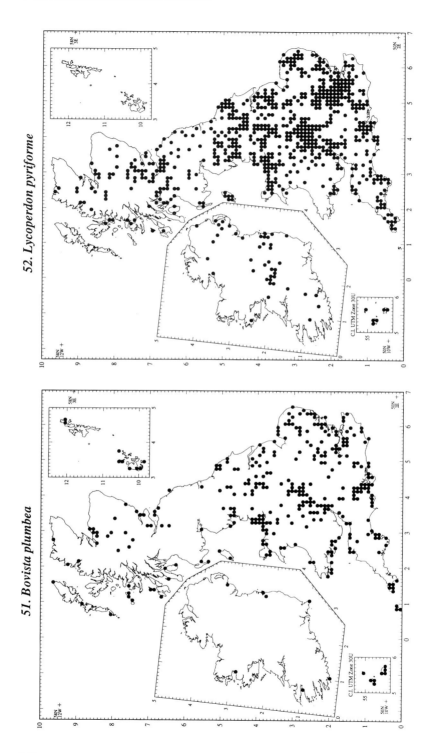

52. Lycoperdon pyriforme

51. Bovista plumbea

British Puffballs, Earthstars and Stinkhorns

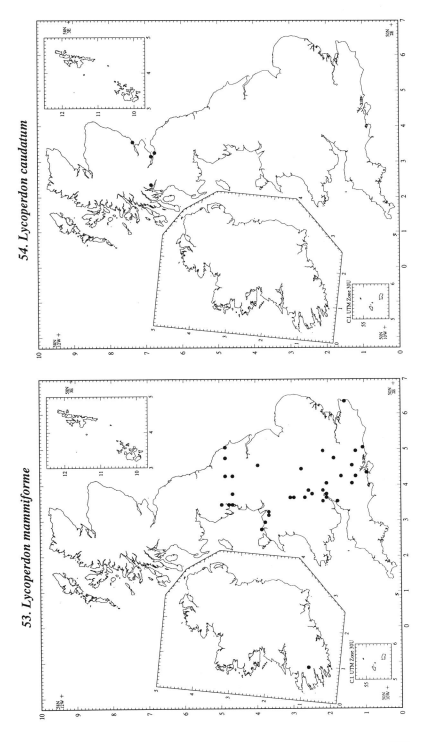

54. Lycoperdon caudatum

53. Lycoperdon mammiforme

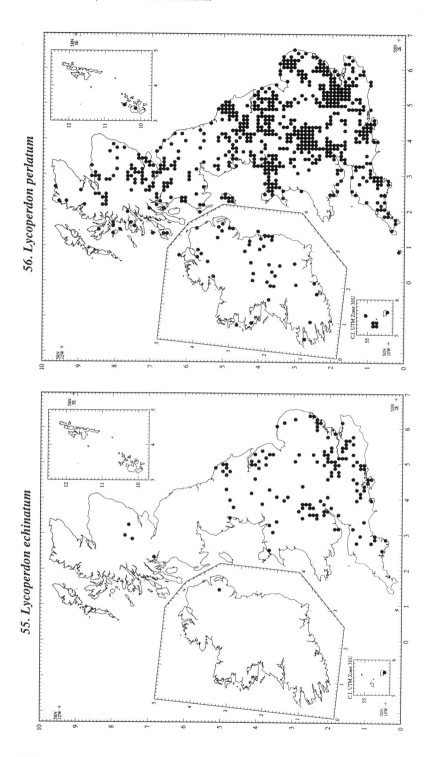

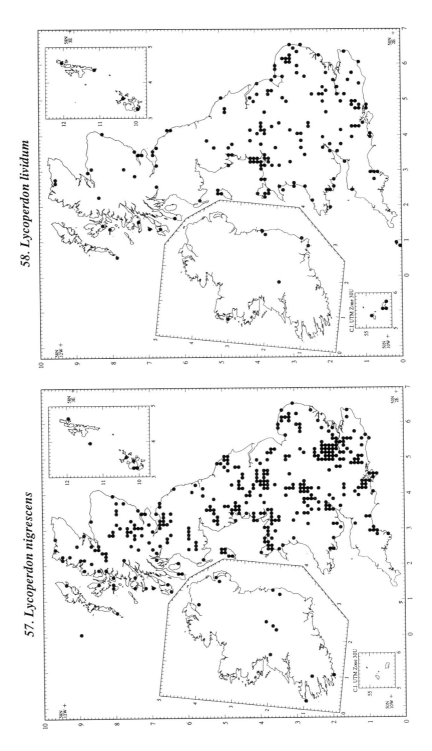

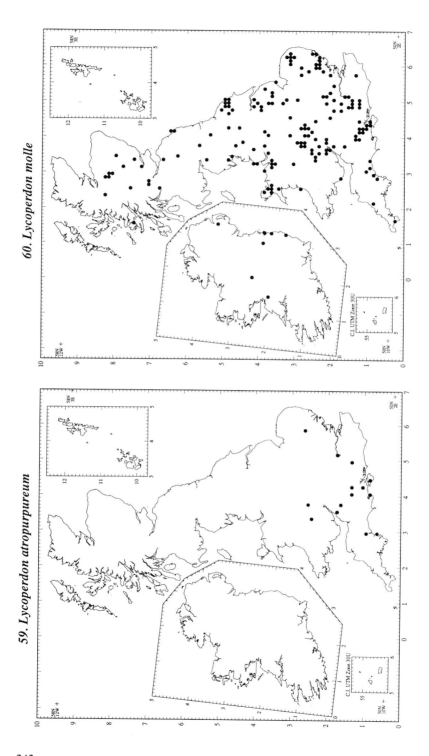

60. Lycoperdon molle

59. Lycoperdon atropurpureum

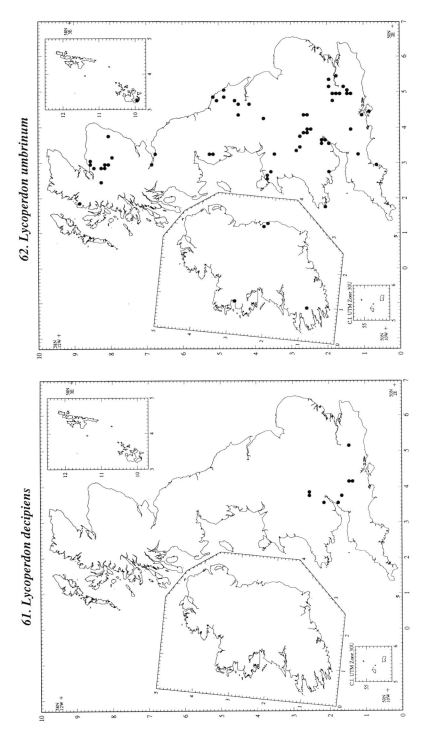

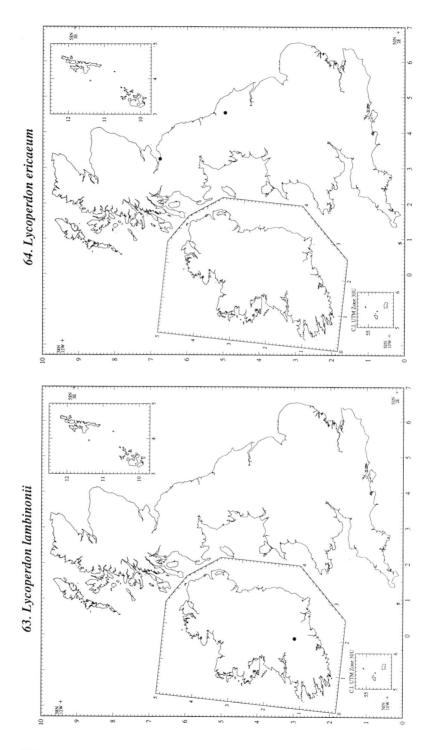

63. Lycoperdon lambinonii

64. Lycoperdon ericaeum

British Puffballs, Earthstars and Stinkhorns

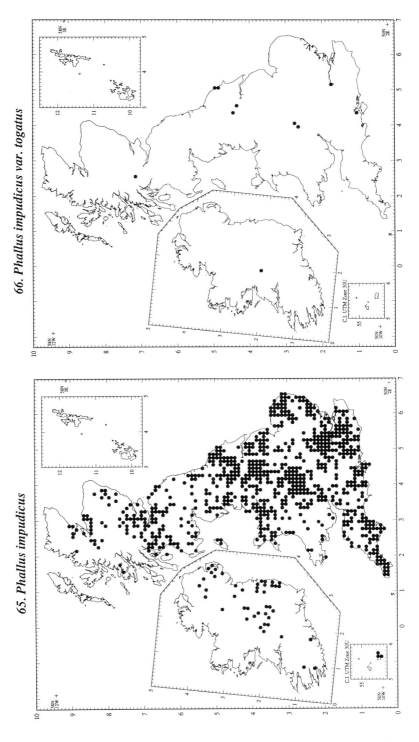

66. Phallus impudicus var. togatus

65. Phallus impudicus

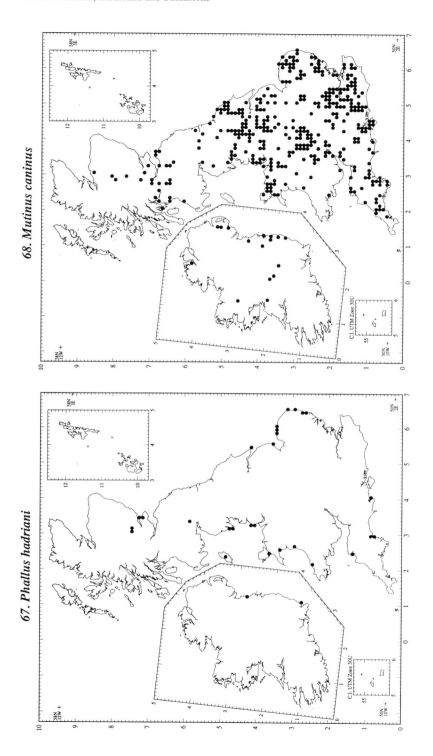

British Puffballs, Earthstars and Stinkhorns

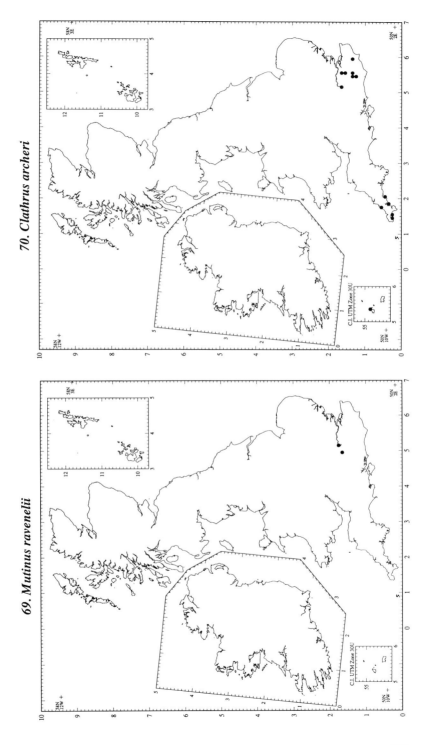

69. Mutinus ravenelii

70. Clathrus archeri

British Puffballs, Earthstars and Stinkhorns

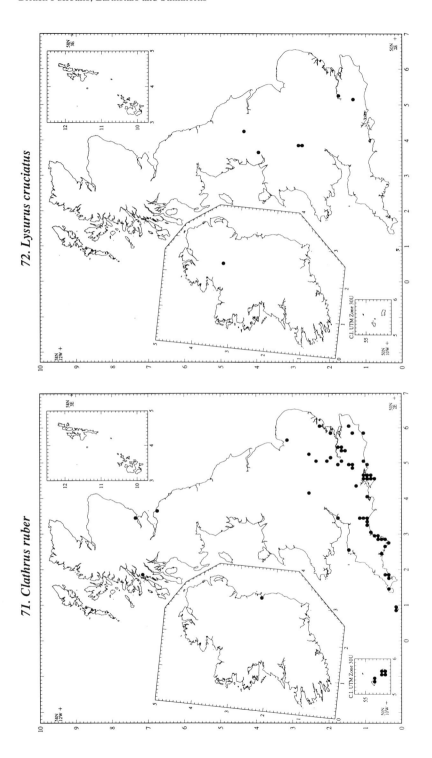

British Puffballs, Earthstars and Stinkhorns

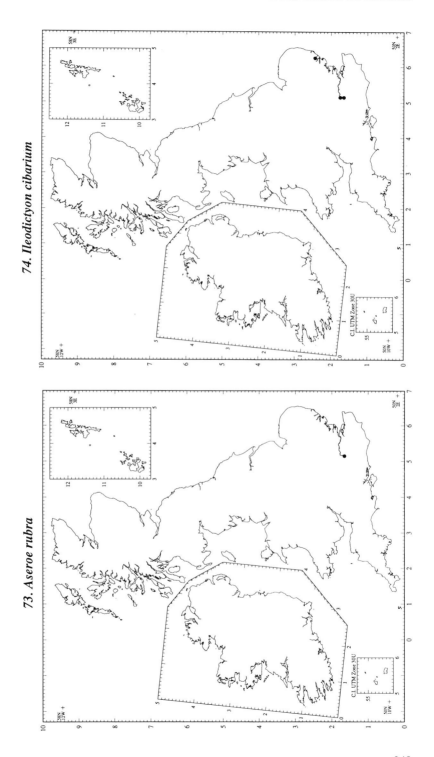

INDEX TO TAXA

Actigea 22
Agaricales 3, 5, 8, 191
Anthurus 181
    archeri 182
    aseroeformis 182
    cruciatus 186
Arched Earthstar 79, **86**
Aseroe 188
    actinobola 188
    hookeri 188
    rubra 2, 10, 180, 182, **188**
    viridis 188
Aserophallus 181
    cruciatus 186
Astraeaceae 6, 20, **38**
Astraeus 3, 6, 8, **38**
    hygrometricus 17, 38, **40**
    koreanus 39
    pteridis 39
    stellatus 40
Atractiella solani 1
Atractiellales 1

Barometer Earthstars 20, **40**
Basket Fungus 180, **190**
Battarraea 5, 16, **52**
    phalloides 10, **53**, 54
    stevenii 53, 54
Battarraeaceae 42, **52**
Battarraeoides 52
Beaked Earthstar 72, **90**
Beard Truffles 191
Berkeley's Earthstar 78, **84**
Bird's Nest Fungi 5, 6, 8, 19, 56, **57**
Blackish Puffball 143, **154**
Blushing Beard Truffle **197**
Bog Mycocalia **67**
Boletales 8, **191**
Bovista 5, 8, 14, 115, **128**, 142
    aestivalis 128, **130**, 132, 156
    colorata 130
    dermoxantha 128, **132**, 136
    gigantea 116
    graveolens 16, 129, 138
    limosa 128, 132, **134**
    nigrescens 17, 129, **138**
    paludosa 128, **134**
    plumbea 119, 129, **140**
    pusilla 132
    pusilliformis 130
    tomentosa 129, 136
Bovistella 5, 115, **120**
    paludosa 134
    radicata 16, **120**
    sinensis 16
Bovists **76**

Broome's Slime Truffle 194
Brown Bovist 129, **138**

Cage-fungi 19, 170, 180, **181**
Calostoma 42
Calostomataceae 42
Calvatia 6, 15, 114, 115, **116**, 122
    caelata 124
    candida 122
    craniiformis 16
    depressa 116
    excipuliformis 126
    fragilis 122
    gigantea 3, 6, 8, 17, **116**, 122
    utriformis 17, 124
Cannon Fungus 6, 8, 19
Caromyxa 176
Carpobolus stellatus 74
Carrot Red Truffle **192**
Chionosphaeraceae
Chlamydopus 42
Clathraceae 5, 7, 8, 19, 170, **180**
Clathrus 180, **181**, 188
    archeri 2, 10, 181, **182**
    cancellatus 184
    cibarius 190
    ruber 16, 180, 181, **184**
Collared Earthstar 79, **108**
Common Beard Truffle **197**
Common Earthball 23, **26**, 32
Common Mycocalia **71**
Common Nut Truffle **193**
Common Puffball 143, **152**
Common Stinkhorn 171, **172**
Conifer Puffball **166**
Cortinariales 8, 191, **192,**
Corynites 176, 178
    ravenelii 178
Cracking Nut Truffle **193**
Crowned Earthstar 79, **104**
Crucibulum 56, **64**
    laeve 62, **64**
    vulgare 64
Cyathus 16, 56, **57**
    crucibulum 64
    denudatus 66, 71
    olla 57, 60, **62**
        var. agrestis 62
        forma anglicus 62
    stercoreus 11, 57, **60**
    striatus 17, 57, **58**, 64
    verniciosus 62
Cynophallus 176

Daisy Earthstar 79, **92**
Dark-spored Puffball 143, 158

251

British Puffballs, Earthstars and Stinkhorns

Deceiving Bovist 128, **130**
Devil's Fingers 181, **182**
Dictyocephalus 42
Dictyophora 17, 171
  duplicata 172
Disciseda 114
Dog Stinkhorn 171, 176, **177**
Dune Stinkhorn 171
Dung Bird's Nest 57
Durieu's Mycocalia **72**
Dwarf Bovist 128, **132**
Dwarf Earthstar 78, **86**
Dyeball 10, **20**, 21

Earthballs 3, 10, 20, 21, **22**
Earthstars 3, 19, 76, **77**
Elasmomycetaceae 5, 191, **192**
Elegant Earthstar 79, **80**
Endogonopsis 38

Fen Bovist 128, **134**
Field Bird's Nest 57
Field Earthstar 79, **82**
Flaky Puffball 142, **146**
Flask-shaped Earthstar 19, **110**
Flesh-Pink Truffle **191**
Fluted Bird's Nest 57, **58**
Four-rayed Earthstar 79, **98**

Gastrosporium simplex 7
Gautieriaceae 191
Gautieria morchelliformis 191
Geaster 77
  stellatus 38
Geastraceae 20, 76, **76**
Geastrum 5, 13, 38, **77**
  ambiguum 83
  asperum 82
  atratum 104
  berkeleyi 78, 83, **84**
  bryantii 88
  campestre 78, **82**, 83
  cesatii 100
  coliforme 112
  columnatum 112
  corollinum 79, 92, **106**
  coronatum 78, 79, 100, 102, **104**
  cryptorhynchum 108
  delicatum 92
  elegans 78, **80**, 86
  fenestratum 96
  fibrillosum 40
  fimbriatum 79, 80, **94**, 95, 102
  floriforme 79, **92**, 106
  fornicatum 14, 79, **96**, 98
  granulosum 100
  hollosii 84
  hygrometricum 40
  kalchbrenneri 108

lageniforme 70, 109, **110**
limbatum 104
mammosum 106
marchicum 96
marginatum 100
michelianum 109
minimum 79, 98, **100**, 104
multifidum 78
nanum 86
orientale 88
pazschkeanum 92
pectinatum 78, 86, 88, **90**
pillotii 108
plicatum 90
pseudomammosum 82
pseudostriatum 84
quadrifidum 79, 96, **98**, 100
recolligens 106
rufescens 78, 79, 94, 95, **102**, 104
schaefferi 102
schmidelii 78, 80, **86**, 90
sessile 94
sibiricum 92
striatum 78, **88**, 90
tenuipes 90
triplex 79, **108**, 110
tunicatum 94, 109
vulgatum 102
Giant Puffball 3, 6, 8, 16, **116**
Glischroderma cinctum 20
Glischrodermataceae 20
Granularia 68
  denudata 71
  duriaeana 72
Grassland Puffball 143, **156**
Grey Stinkhorn Truffle **193**
Gymnomyces xanthosporus 192

Handkea 5, 114, 116, **122**
  excipuliformis 122, **126**
  utriformis 122, **124**
Heath Puffball 143, **168**
Hedgehog Puffball 142, **150**
Heterogastridiales 1
Heterogastridium 1
Honeycomb Truffle 21, **191**
Hydnangiaceae 191
Hydnangium carneum 191
Hymenangiaceae 194
Hymenangium album 194
Hymenogaster 3, **192**
  arenarius 195
  citrinus 194
  griseus 196
  hessei 194
  luteus 194
  muticus 195
  olivaceus 195
  sulcatus 195

Index to Taxa

tener 195
thwaitesii 195
vulgaris 195
Hymenogasteraceae 194
Hymenogastrales 7
Hymenophallus 171
  togatus 172
Hysterangiaceae 170, 180, **193**
Hysterangium 170, **196**
  coriaceum 196
  nephriticum 196
  thwaitesii 196

Ileodictyon 16, **190**
  cibarium 190
Ithyphallus 171
  impudicus 172
    var. iosmus 174

Kirchbaumia imperialis 174

Langermannia 116, 122
  gigantea 116
Lasiosphaera 116
  gigantea 116
Lead-grey Bovist 129, **140**
Leaf-litter Truffles **192**
Least Bovist 128, **136**
Lemon-coloured Nut Truffle **194**
Leopard-Spotted Earthball 34
Leucogaster 21
  nudus 191
Leucogastraceae 191
Linderia 181
Lizard's Claw 180, **186**
Lycoperdaceae 20, 96, **114**
Lycoperdales 1, 3, 5, 7, 8, 19, **76**, 114
Lycoperdon 2, 5, 6, 8, 13, 14, 15, 16, 22,
  114, 115, 128, **142**
  aestivale 130
  atropurpureum 143, **158**, 162, 164
  aurantium 26
  bovista 124
  caelatum 124
  candidum 148
  carpobolus 74
  caudatum 134, 142, **148**
  cepaeforme 130
  coliforme 112
  coloratum 130
  corollinum 106
  coronatum 98
  craniforme 116
  cruciatum 146
  decipiens 143, 158, **162**, 164
  depressum 118
  dermoxanthum 132
  elatatum 126
  echinatum 142, **150**

ericaeum 143, **168**
  var. subareolatum 168
ericetorum 132
estonicum 164
excipuliforme 126
fenestratum 96
foetidum 154
fornicatum 96
frigidum 143, 160
furfuracea 130
gemmatum var. perlatum 152
giganteum 116
hoylei 150, 154
hyemale 118
lambinonii 143, **166**
limosum 134
lividum 143, **156**
mammiforme 142, **146**, 150
mammosum 46
mauryanum 158
molle 126, 143, 146, 158, **160**, 162,
  164, 166
muscorum 168
nigrescens 143, 153, **154**
niveum 160
norvegicum 143, 148, 152
pedunculatum 46
perlatum 143, **152**, 160
  var. nigrescens 154
phalloides 53
polymorphum 130
polyrhizum 28
pratense 118
pyriforme 9, 10, 142, **144**
  var. excipuliforme 144
    tesselatum 144
radicatum 120
recolligens 106
sessile 94
sinclairii 124
spadiceum 156
stellatum 40
tessulatum 26
umbrinum 143, 156, 160, **164**, 166
utriforme 124
velatum 146
verrucosum 32
Lysurus 186, 188
  archeri 182
  cruciatus 180, **186**
  gardneri 186
  mokusin 16, 17

Many-rooted Earthball 23, **28**
Melanogastraceae 191
Melanogastrales 5, 7, 8
Melanogaster 191, **194**
  ambiguus 194
  broomeianus 194

253

intermedius 194
Mesophelliaceae 77
Morganella subincarnata 16
Mosaic Puffball 122
Milk Cap Truffles 5, 191
Mutinus 6, 8, 171, **176**
  bambusinus 178
  caninus 176, **177**, 178
  ravenelii 176, **178**
Mycenastraceae 114
Mycenastrum corium 114
Mycoaureola dilseae 1
Mycocalia 3, 10, 56, **68**
  denudata 68, 70, **71**
  duriaeana 68, 70, **72**
  minutissima 68, **70**
  sphagneti 68, **69**
Myriostoma 6, 8, 10, 38, 77, **112**
  anglicum 112
  coliforme 7, **112**

Nia vibrissa 1
Nidularia 56, **66**, 68
  arundinacea 71
  berkeleyi 66
  campanulata 62
  castanea 72
  confluens 66
  deformis 66
    var. **confluens 66**
      radicata 66
  denudata 68, 71
  duriaeana 72
  farcta 66
  fusispora 71
  hirsuta 58
  laevis 64
  minutissima 70
  pisiformis 66
    var. broomei 66
  stercorea 60
  striata 58
Nidulariaceae 6, 8, 19, **56**
Nidulariales 1, 5, 7, 8, 19, **56**
Nut Truffles **192**

Octavianina asterosperma 197
Octavianinaceae 192, **197**
Olive Nut Truffle **193**
Onion Earthball 23, **30**

Pachnocybaceae 1
Pachnocybe ferruginea 1
Parasitic Boletus 26
Pedicelled Puffball 142, **148**
Pepper Pot 6, 8, 77, **112**
Pestle-shaped Puffball 122, **126**
Peziza striatus 58
Phallaceae 7, 19, **170**, 180

Phallales 6, 8, 13, **170**, 191
Phallus 6, 8, 13, **171**
  caninus 177
  duplicatus 172
  foetidus 172
  hadriani 10, 171, **174**
  imperiale 174
  impudicus 171, **172**
    var. **impudicus** 171, **172**
      pseudoduplicatus 172
      togatus 171, **172**
  iosmus 174
  rubicundus 17
Phloeogena faginea 1
Phlyctospora 22
  fusca 24
Pisolithus 21, **36**, 39
  arhizus 10, 17, **36**
  tinctorius 16, 36
Podaxis 3
Polysaccum pisocarpium 36
  tinctorium 36
Polystoma 112
  coliforme 112
Pompholyx 22
Potato Earthball 22, **24**
Protubera 180
Puffballs 12, 20, 76
Pyrenogaster 77

Queletia 42, **50**
  andina 50
  mirabilis 50
Quélet's Stalk Puffball 42, **50**

Radiigera 76
Red Cage Fungus 181, **184**
Red Stinkhorn 176, **178**
Reticulated Beard Truffle **195**
Rhizopogonaceae 191
Rhizopogon 191, **197**
  luteolus 197
  reticulatus 197
  roseolus 197
  vulgaris 197
Rosy Earthstar 79, **102**
Russulales 191

Sandy Stilt Puffball **53**
Scaly Earthball 23, **24**
Scaly Stalk Puffball 43, **48**
Schizostoma 42
Sclerangium 22
  polyrhizum 28
Scleroderma 10, 13, 15, 21, **22**, 39
  areolatum 34
  arhizum 36
  aurantium 26
  bovista 22, **24**

cepa 16, 23, **30**
cepioides 30
citrinum 23, **26**, 28, 32
fuscum 24
geaster 28
lycoperdoides 34
polyrhizum 23, **28**
texense 24
verrucosum 22, 23, 24, 30, **32**, 34
  var. bovista 23
violascens 34
vulgare 26
  var. spadiceum 30
Sclerodermataceae 20, **21**
Sclerodermatales 1, 3, 7, 8, 19, **20**
Sclerogaster compactus 197
Sessile Earthstar 79, **94**
Slime Truffles 4, **191**
Soft-spined Puffball 143, **160**
Sphaerobolaceae 19, 56, **74**
Sphaerobolus 6, 8, **74**
  carpobolus 74
  dentatus 74
  stellatus 74
  terrestris 74
Stalk Puffballs 3, 19, 42, **43**
Starfish Fungus 180, **188**
Stella 22
  americana 28
Stephanospora caroticolor 192
Stephanosporaceae 192
Steppe Puffball 143, **162**
Steppe Truffle **192**
Stereales 192
Stilbum vulgare 1
Stilt Puffball 4, 19, 42
Stinkhorn Truffles 170, **191**
Stinkhorns 3, 12, 19, 170, **171**
Stinking Slime Truffle **194**
Striated Earthstar 78, **88**
Stump Puffball 142, **144**

Thin Nut Truffle **193**
Thwaites's Nut Truffle **196**
Thwaites's Stinkhorn Truffle **194**
Tiny Earthstar 79, **100**
Tiny Mycocalia **70**
Toothed Phallus 174
Tremellogaster 21
Tuber 22
Tulasnodea 43
  mammosa 46
Tulostoma 6, 16, 42, **43**, 50
  brevipes 46
  brumale 43, **46**, 48, 136
    var. bataviense 48
  fuscoviolaceostipitatum 46
  mammosa 46
  melanocyclum 43, **48**

niveum 10, 43, **44**
  pedunculatum 46
Tulostomataceae 19, 42
Tulostomatales 1, 3, 6, 7, 8, 19, **42**
Tylostoma 43

Umber-brown Puffball 143, **164**

Vascellum 6, 16, 115, **118**
  depressum 118
  pratense 118, 140
Veligaster 21

Wakefieldia macrospora 197
Weathered Earthstar 79, **106**
White-egg Bird's Nest **64**
White Stalk Puffball 43, **44**
Winter Stalk Puffball 43, **46**

Xerocomus parasiticus 26

Yellow Beard Truffle **197**
Yellow Nut Truffle **194**

Zelleromyces stephensii 192